# Tests and Worksheets

**SAXON Math™**
**HOMESCHOOL**
**7/6**

**Stephen Hake**

**John Saxon**

**SAXON™**
PUBLISHERS

Saxon Publishers gratefully acknowledges the contributions of the following individuals in the completion of this project:

**Authors:** Stephen Hake, John Saxon

**Editorial:** Chris Braun, Bo Björn Johnson, Dana Nixon, Matt Maloney, Brian E. Rice

**Editorial Support Services:** Christopher Davey, Jay Allman, Shelley Turner, Jean Van Vleck, Darlene Terry

**Production:** Alicia Britt, Karen Hammond, Donna Jarrel, Brenda Lopez, Adriana Maxwell, Cristi D. Whiddon

**Project Management:** Angela Johnson, Becky Cavnar

Printed in the United States of America

ISBN: 1-59141-323-0

4 5 6 7 8   862   12 11 10 09 08 07

# CONTENTS

# Introduction

*Saxon Math 7/6—Homeschool Tests and Worksheets* contains Facts Practice Tests, Activity Sheets, tests, and recording forms. Brief descriptions of these components are provided below, and additional information can be found on the pages that introduce each section. Solutions to the Facts Practice Tests, Activity Sheets, and tests are located in the *Saxon Math 7/6—Homeschool Solutions Manual.* For a complete overview of the philosophy and implementation of Saxon Math™, please refer to the preface of the *Saxon Math 7/6—Homeschool* textbook.

## About the Facts Practice Tests

Facts Practice Tests are an essential and integral part of Saxon Math™. Mastery of basic facts frees your student to focus on procedures and concepts rather than computation. Employing memory to recall frequently encountered facts permits students to bring higher-level thinking skills to bear when solving problems.

Facts Practice Tests should be administered as prescribed at the beginning of each lesson or test. Sufficient copies of the Facts Practice Tests for one student are supplied, in the order needed, with the corresponding lesson or test clearly indicated at the top of the page. Limit student work on these tests to five minutes or less. Your student should keep track of his or her times and scores and get progressively faster and more accurate as the course continues.

## About the Activity Sheets

Selected lessons and investigations in the student textbook present content through activities. These activities often require the use of worksheets called Activity Sheets, which are provided in this workbook in the quantities needed by one student.

## About the Tests

The tests are designed to reward your student and to provide you with diagnostic information. Every lesson in the student textbook culminates with a cumulative mixed practice, so the tests are cumulative as well. By allowing your student to display his or her skills, the tests build confidence and motivation for continued learning. The cumulative nature of Saxon tests also gives your student an incentive to master skills and concepts that might otherwise be learned for just one test.

All the tests needed for one student are provided in this workbook. The testing schedule is printed on the page immediately preceding the first test. Administering the tests according to the schedule is essential. Each test is written to follow a specific five-lesson interval in the textbook. Following the schedule allows your student to gain sufficient practice on new topics before being tested over them.

## About the Recording Forms

The last section of this book contains five optional recording forms. Three of the forms provide an organized framework for your student to record his or her work on the daily lessons, Mixed Practices, and tests. Two of the forms help track and analyze your student's performance on his or her assignments. All five of the recording forms may be photocopied as needed.

# *Facts Practice Tests*
# *and Activity Sheets*

This section contains the Facts Practice Tests and Activity Sheets, which are sequenced in the order of their use in *Saxon Math 7/6—Homeschool.* Sufficient copies for one student are provided.

**Facts Practice Tests**
Rapid and accurate recall of basic facts and skills dramatically increases students' mathematical abilities. To that end we have provided the Facts Practice Tests. Begin each lesson with the Facts Practice Test suggested in the Warm-Up, limiting the time to five minutes or less. Your student should work independently and rapidly during the Facts Practice Tests, trying to improve on previous performances in both speed and accuracy.

Each Facts Practice Test contains a line for your student to record his or her time. Timing the student is motivating. Striving to improve speed helps students automate skills and offers the additional benefit of an up-tempo atmosphere to start the lesson. Time invested in practicing basic facts is repaid in your student's ability to work faster.

After each Facts Practice Test, quickly read aloud the answers from the *Saxon Math 7/6—Homeschool Solutions Manual* as your student checks his or her work. If your student made any errors or was unable to finish within the allotted time, he or she should correct the errors or complete the problems as part of the day's assignment. You might wish to have your student track Facts Practice scores and times on Recording Form A, which is found in this workbook.

On test day the student should be held accountable for mastering the content of recent Facts Practice Tests. Hence, each test identifies a Facts Practice Test to be taken on that day. Allow five minutes on test days for the student to complete the Facts Practice Test before beginning the cumulative test.

**Activity Sheets**
Activity Sheets are referenced in certain lessons and investigations of *Saxon Math 7/6—Homeschool.* Students should refer to the textbook for detailed instructions on using the Activity Sheets. The fraction manipulatives (on Activity Sheets 3–7) may be color-coded with colored pencils or markers before they are cut out.

## B    100 Addition Facts
*For use with Lesson 3*

Name _____

Time _____

Add.

| | | | | | | | | | |
|---|---|---|---|---|---|---|---|---|---|
| 3 <br> + 2 | 8 <br> + 3 | 2 <br> + 1 | 5 <br> + 6 | 2 <br> + 9 | 4 <br> + 8 | 8 <br> + 0 | 3 <br> + 9 | 1 <br> + 0 | 6 <br> + 3 |
| 7 <br> + 3 | 1 <br> + 6 | 4 <br> + 7 | 0 <br> + 3 | 6 <br> + 4 | 5 <br> + 5 | 3 <br> + 1 | 7 <br> + 2 | 8 <br> + 5 | 2 <br> + 5 |
| 4 <br> + 0 | 5 <br> + 7 | 1 <br> + 1 | 5 <br> + 4 | 2 <br> + 8 | 7 <br> + 1 | 4 <br> + 6 | 0 <br> + 2 | 6 <br> + 5 | 4 <br> + 9 |
| 8 <br> + 6 | 0 <br> + 4 | 5 <br> + 8 | 7 <br> + 4 | 1 <br> + 7 | 6 <br> + 6 | 4 <br> + 1 | 8 <br> + 2 | 2 <br> + 4 | 6 <br> + 0 |
| 9 <br> + 1 | 8 <br> + 8 | 2 <br> + 2 | 4 <br> + 5 | 6 <br> + 2 | 0 <br> + 0 | 5 <br> + 9 | 3 <br> + 3 | 8 <br> + 1 | 2 <br> + 7 |
| 4 <br> + 4 | 7 <br> + 5 | 0 <br> + 1 | 8 <br> + 7 | 3 <br> + 4 | 7 <br> + 9 | 1 <br> + 2 | 6 <br> + 7 | 0 <br> + 8 | 9 <br> + 2 |
| 0 <br> + 9 | 8 <br> + 9 | 7 <br> + 6 | 1 <br> + 3 | 6 <br> + 8 | 2 <br> + 0 | 8 <br> + 4 | 3 <br> + 5 | 9 <br> + 8 | 5 <br> + 0 |
| 9 <br> + 3 | 2 <br> + 6 | 3 <br> + 0 | 6 <br> + 1 | 3 <br> + 6 | 5 <br> + 2 | 0 <br> + 5 | 6 <br> + 9 | 1 <br> + 8 | 9 <br> + 6 |
| 4 <br> + 3 | 9 <br> + 9 | 0 <br> + 7 | 9 <br> + 4 | 7 <br> + 7 | 1 <br> + 4 | 3 <br> + 7 | 7 <br> + 0 | 2 <br> + 3 | 5 <br> + 1 |
| 9 <br> + 5 | 1 <br> + 5 | 9 <br> + 0 | 3 <br> + 8 | 1 <br> + 9 | 5 <br> + 3 | 4 <br> + 2 | 9 <br> + 7 | 0 <br> + 6 | 7 <br> + 8 |

**64 Addition Facts**
*For use with Lesson 4*

Name _____

Time _____

Add.

| | | | | | | | |
|---|---|---|---|---|---|---|---|
| 7<br>+ 2 | 9<br>+ 4 | 2<br>+ 8 | 6<br>+ 5 | 4<br>+ 4 | 3<br>+ 9 | 8<br>+ 4 | 5<br>+ 7 |
| 9<br>+ 7 | 4<br>+ 7 | 7<br>+ 5 | 5<br>+ 4 | 3<br>+ 4 | 6<br>+ 8 | 2<br>+ 5 | 8<br>+ 8 |
| 6<br>+ 3 | 2<br>+ 9 | 7<br>+ 8 | 8<br>+ 3 | 5<br>+ 9 | 3<br>+ 6 | 9<br>+ 9 | 4<br>+ 9 |
| 5<br>+ 8 | 9<br>+ 5 | 4<br>+ 5 | 8<br>+ 6 | 2<br>+ 3 | 6<br>+ 6 | 5<br>+ 2 | 7<br>+ 3 |
| 3<br>+ 8 | 8<br>+ 9 | 2<br>+ 2 | 7<br>+ 6 | 5<br>+ 5 | 6<br>+ 9 | 3<br>+ 7 | 9<br>+ 8 |
| 4<br>+ 2 | 3<br>+ 3 | 6<br>+ 4 | 4<br>+ 8 | 9<br>+ 3 | 2<br>+ 4 | 8<br>+ 5 | 7<br>+ 9 |
| 7<br>+ 4 | 2<br>+ 6 | 5<br>+ 3 | 9<br>+ 6 | 4<br>+ 3 | 6<br>+ 7 | 3<br>+ 2 | 8<br>+ 7 |
| 5<br>+ 6 | 8<br>+ 2 | 3<br>+ 5 | 6<br>+ 2 | 7<br>+ 7 | 4<br>+ 6 | 9<br>+ 2 | 2<br>+ 7 |

# B — 100 Addition Facts

*For use with Lesson 5*

Name _____

Time _____

Add.

| | | | | | | | | | |
|---|---|---|---|---|---|---|---|---|---|
| 3<br>+ 2 | 8<br>+ 3 | 2<br>+ 1 | 5<br>+ 6 | 2<br>+ 9 | 4<br>+ 8 | 8<br>+ 0 | 3<br>+ 9 | 1<br>+ 0 | 6<br>+ 3 |
| 7<br>+ 3 | 1<br>+ 6 | 4<br>+ 7 | 0<br>+ 3 | 6<br>+ 4 | 5<br>+ 5 | 3<br>+ 1 | 7<br>+ 2 | 8<br>+ 5 | 2<br>+ 5 |
| 4<br>+ 0 | 5<br>+ 7 | 1<br>+ 1 | 5<br>+ 4 | 2<br>+ 8 | 7<br>+ 1 | 4<br>+ 6 | 0<br>+ 2 | 6<br>+ 5 | 4<br>+ 9 |
| 8<br>+ 6 | 0<br>+ 4 | 5<br>+ 8 | 7<br>+ 4 | 1<br>+ 7 | 6<br>+ 6 | 4<br>+ 1 | 8<br>+ 2 | 2<br>+ 4 | 6<br>+ 0 |
| 9<br>+ 1 | 8<br>+ 8 | 2<br>+ 2 | 4<br>+ 5 | 6<br>+ 2 | 0<br>+ 0 | 5<br>+ 9 | 3<br>+ 3 | 8<br>+ 1 | 2<br>+ 7 |
| 4<br>+ 4 | 7<br>+ 5 | 0<br>+ 1 | 8<br>+ 7 | 3<br>+ 4 | 7<br>+ 9 | 1<br>+ 2 | 6<br>+ 7 | 0<br>+ 8 | 9<br>+ 2 |
| 0<br>+ 9 | 8<br>+ 9 | 7<br>+ 6 | 1<br>+ 3 | 6<br>+ 8 | 2<br>+ 0 | 8<br>+ 4 | 3<br>+ 5 | 9<br>+ 8 | 5<br>+ 0 |
| 9<br>+ 3 | 2<br>+ 6 | 3<br>+ 0 | 6<br>+ 1 | 3<br>+ 6 | 5<br>+ 2 | 0<br>+ 5 | 6<br>+ 9 | 1<br>+ 8 | 9<br>+ 6 |
| 4<br>+ 3 | 9<br>+ 9 | 0<br>+ 7 | 9<br>+ 4 | 7<br>+ 7 | 1<br>+ 4 | 3<br>+ 7 | 7<br>+ 0 | 2<br>+ 3 | 5<br>+ 1 |
| 9<br>+ 5 | 1<br>+ 5 | 9<br>+ 0 | 3<br>+ 8 | 1<br>+ 9 | 5<br>+ 3 | 4<br>+ 2 | 9<br>+ 7 | 0<br>+ 6 | 7<br>+ 8 |

**C**

## 100 Subtraction Facts

*For use with Lesson 6*

Name _____

Time _____

Subtract.

| | | | | | | | | | |
|---|---|---|---|---|---|---|---|---|---|
| 16 − 9 | 7 − 1 | 18 − 9 | 11 − 3 | 13 − 7 | 8 − 2 | 11 − 5 | 5 − 0 | 17 − 9 | 6 − 1 |
| 10 − 9 | 6 − 2 | 13 − 4 | 4 − 0 | 10 − 5 | 5 − 1 | 10 − 3 | 12 − 6 | 10 − 1 | 6 − 4 |
| 7 − 2 | 14 − 7 | 8 − 1 | 11 − 6 | 3 − 3 | 16 − 7 | 5 − 2 | 12 − 4 | 3 − 0 | 11 − 7 |
| 17 − 8 | 6 − 0 | 10 − 6 | 4 − 1 | 9 − 5 | 9 − 0 | 5 − 4 | 12 − 5 | 4 − 2 | 9 − 3 |
| 12 − 3 | 16 − 8 | 9 − 1 | 15 − 6 | 11 − 4 | 13 − 5 | 1 − 0 | 8 − 5 | 9 − 6 | 11 − 2 |
| 7 − 0 | 10 − 8 | 6 − 3 | 14 − 5 | 3 − 1 | 8 − 6 | 4 − 4 | 11 − 8 | 3 − 2 | 15 − 9 |
| 13 − 8 | 7 − 4 | 10 − 7 | 0 − 0 | 12 − 8 | 5 − 5 | 4 − 3 | 8 − 7 | 7 − 3 | 7 − 6 |
| 5 − 3 | 7 − 5 | 2 − 1 | 6 − 6 | 8 − 4 | 2 − 2 | 13 − 6 | 15 − 8 | 2 − 0 | 13 − 9 |
| 1 − 1 | 11 − 9 | 10 − 4 | 9 − 2 | 14 − 6 | 8 − 0 | 9 − 4 | 10 − 2 | 6 − 5 | 8 − 3 |
| 7 − 7 | 14 − 8 | 12 − 9 | 9 − 8 | 12 − 7 | 9 − 9 | 15 − 7 | 8 − 8 | 14 − 9 | 9 − 7 |

# C

## 100 Subtraction Facts
*For use with Lesson 7*

Name _____

Time _____

Subtract.

| | | | | | | | | | |
|---|---|---|---|---|---|---|---|---|---|
| 16<br>− 9 | 7<br>− 1 | 18<br>− 9 | 11<br>− 3 | 13<br>− 7 | 8<br>− 2 | 11<br>− 5 | 5<br>− 0 | 17<br>− 9 | 6<br>− 1 |
| 10<br>− 9 | 6<br>− 2 | 13<br>− 4 | 4<br>− 0 | 10<br>− 5 | 5<br>− 1 | 10<br>− 3 | 12<br>− 6 | 10<br>− 1 | 6<br>− 4 |
| 7<br>− 2 | 14<br>− 7 | 8<br>− 1 | 11<br>− 6 | 3<br>− 3 | 16<br>− 7 | 5<br>− 2 | 12<br>− 4 | 3<br>− 0 | 11<br>− 7 |
| 17<br>− 8 | 6<br>− 0 | 10<br>− 6 | 4<br>− 1 | 9<br>− 5 | 9<br>− 0 | 5<br>− 4 | 12<br>− 5 | 4<br>− 2 | 9<br>− 3 |
| 12<br>− 3 | 16<br>− 8 | 9<br>− 1 | 15<br>− 6 | 11<br>− 4 | 13<br>− 5 | 1<br>− 0 | 8<br>− 5 | 9<br>− 6 | 11<br>− 2 |
| 7<br>− 0 | 10<br>− 8 | 6<br>− 3 | 14<br>− 5 | 3<br>− 1 | 8<br>− 6 | 4<br>− 4 | 11<br>− 8 | 3<br>− 2 | 15<br>− 9 |
| 13<br>− 8 | 7<br>− 4 | 10<br>− 7 | 0<br>− 0 | 12<br>− 8 | 5<br>− 5 | 4<br>− 3 | 8<br>− 7 | 7<br>− 3 | 7<br>− 6 |
| 5<br>− 3 | 7<br>− 5 | 2<br>− 1 | 6<br>− 6 | 8<br>− 4 | 2<br>− 2 | 13<br>− 6 | 15<br>− 8 | 2<br>− 0 | 13<br>− 9 |
| 1<br>− 1 | 11<br>− 9 | 10<br>− 4 | 9<br>− 2 | 14<br>− 6 | 8<br>− 0 | 9<br>− 4 | 10<br>− 2 | 6<br>− 5 | 8<br>− 3 |
| 7<br>− 7 | 14<br>− 8 | 12<br>− 9 | 9<br>− 8 | 12<br>− 7 | 9<br>− 9 | 15<br>− 7 | 8<br>− 8 | 14<br>− 9 | 9<br>− 7 |

## A | 64 Addition Facts
*For use with Lesson 8*

Name _____

Time _____

Add.

| | | | | | | | |
|---|---|---|---|---|---|---|---|
| 7<br>+ 2 | 9<br>+ 4 | 2<br>+ 8 | 6<br>+ 5 | 4<br>+ 4 | 3<br>+ 9 | 8<br>+ 4 | 5<br>+ 7 |
| 9<br>+ 7 | 4<br>+ 7 | 7<br>+ 5 | 5<br>+ 4 | 3<br>+ 4 | 6<br>+ 8 | 2<br>+ 5 | 8<br>+ 8 |
| 6<br>+ 3 | 2<br>+ 9 | 7<br>+ 8 | 8<br>+ 3 | 5<br>+ 9 | 3<br>+ 6 | 9<br>+ 9 | 4<br>+ 9 |
| 5<br>+ 8 | 9<br>+ 5 | 4<br>+ 5 | 8<br>+ 6 | 2<br>+ 3 | 6<br>+ 6 | 5<br>+ 2 | 7<br>+ 3 |
| 3<br>+ 8 | 8<br>+ 9 | 2<br>+ 2 | 7<br>+ 6 | 5<br>+ 5 | 6<br>+ 9 | 3<br>+ 7 | 9<br>+ 8 |
| 4<br>+ 2 | 3<br>+ 3 | 6<br>+ 4 | 4<br>+ 8 | 9<br>+ 3 | 2<br>+ 4 | 8<br>+ 5 | 7<br>+ 9 |
| 7<br>+ 4 | 2<br>+ 6 | 5<br>+ 3 | 9<br>+ 6 | 4<br>+ 3 | 6<br>+ 7 | 3<br>+ 2 | 8<br>+ 7 |
| 5<br>+ 6 | 8<br>+ 2 | 3<br>+ 5 | 6<br>+ 2 | 7<br>+ 7 | 4<br>+ 6 | 9<br>+ 2 | 2<br>+ 7 |

## 100 Subtraction Facts

*For use with Lesson 9*

Name _____

Time _____

Subtract.

| | | | | | | | | | |
|---|---|---|---|---|---|---|---|---|---|
| 16<br>− 9 | 7<br>− 1 | 18<br>− 9 | 11<br>− 3 | 13<br>− 7 | 8<br>− 2 | 11<br>− 5 | 5<br>− 0 | 17<br>− 9 | 6<br>− 1 |
| 10<br>− 9 | 6<br>− 2 | 13<br>− 4 | 4<br>− 0 | 10<br>− 5 | 5<br>− 1 | 10<br>− 3 | 12<br>− 6 | 10<br>− 1 | 6<br>− 4 |
| 7<br>− 2 | 14<br>− 7 | 8<br>− 1 | 11<br>− 6 | 3<br>− 3 | 16<br>− 7 | 5<br>− 2 | 12<br>− 4 | 3<br>− 0 | 11<br>− 7 |
| 17<br>− 8 | 6<br>− 0 | 10<br>− 6 | 4<br>− 1 | 9<br>− 5 | 9<br>− 0 | 5<br>− 4 | 12<br>− 5 | 4<br>− 2 | 9<br>− 3 |
| 12<br>− 3 | 16<br>− 8 | 9<br>− 1 | 15<br>− 6 | 11<br>− 4 | 13<br>− 5 | 1<br>− 0 | 8<br>− 5 | 9<br>− 6 | 11<br>− 2 |
| 7<br>− 0 | 10<br>− 8 | 6<br>− 3 | 14<br>− 5 | 3<br>− 1 | 8<br>− 6 | 4<br>− 4 | 11<br>− 8 | 3<br>− 2 | 15<br>− 9 |
| 13<br>− 8 | 7<br>− 4 | 10<br>− 7 | 0<br>− 0 | 12<br>− 8 | 5<br>− 5 | 4<br>− 3 | 8<br>− 7 | 7<br>− 3 | 7<br>− 6 |
| 5<br>− 3 | 7<br>− 5 | 2<br>− 1 | 6<br>− 6 | 8<br>− 4 | 2<br>− 2 | 13<br>− 6 | 15<br>− 8 | 2<br>− 0 | 13<br>− 9 |
| 1<br>− 1 | 11<br>− 9 | 10<br>− 4 | 9<br>− 2 | 14<br>− 6 | 8<br>− 0 | 9<br>− 4 | 10<br>− 2 | 6<br>− 5 | 8<br>− 3 |
| 7<br>− 7 | 14<br>− 8 | 12<br>− 9 | 9<br>− 8 | 12<br>− 7 | 9<br>− 9 | 15<br>− 7 | 8<br>− 8 | 14<br>− 9 | 9<br>− 7 |

## C

### 100 Subtraction Facts
*For use with Lesson 10*

Name _____

Time _____

Subtract.

| | | | | | | | | | |
|---|---|---|---|---|---|---|---|---|---|
| 16<br>− 9 | 7<br>− 1 | 18<br>− 9 | 11<br>− 3 | 13<br>− 7 | 8<br>− 2 | 11<br>− 5 | 5<br>− 0 | 17<br>− 9 | 6<br>− 1 |
| 10<br>− 9 | 6<br>− 2 | 13<br>− 4 | 4<br>− 0 | 10<br>− 5 | 5<br>− 1 | 10<br>− 3 | 12<br>− 6 | 10<br>− 1 | 6<br>− 4 |
| 7<br>− 2 | 14<br>− 7 | 8<br>− 1 | 11<br>− 6 | 3<br>− 3 | 16<br>− 7 | 5<br>− 2 | 12<br>− 4 | 3<br>− 0 | 11<br>− 7 |
| 17<br>− 8 | 6<br>− 0 | 10<br>− 6 | 4<br>− 1 | 9<br>− 5 | 9<br>− 0 | 5<br>− 4 | 12<br>− 5 | 4<br>− 2 | 9<br>− 3 |
| 12<br>− 3 | 16<br>− 8 | 9<br>− 1 | 15<br>− 6 | 11<br>− 4 | 13<br>− 5 | 1<br>− 0 | 8<br>− 5 | 9<br>− 6 | 11<br>− 2 |
| 7<br>− 0 | 10<br>− 8 | 6<br>− 3 | 14<br>− 5 | 3<br>− 1 | 8<br>− 6 | 4<br>− 4 | 11<br>− 8 | 3<br>− 2 | 15<br>− 9 |
| 13<br>− 8 | 7<br>− 4 | 10<br>− 7 | 0<br>− 0 | 12<br>− 8 | 5<br>− 5 | 4<br>− 3 | 8<br>− 7 | 7<br>− 3 | 7<br>− 6 |
| 5<br>− 3 | 7<br>− 5 | 2<br>− 1 | 6<br>− 6 | 8<br>− 4 | 2<br>− 2 | 13<br>− 6 | 15<br>− 8 | 2<br>− 0 | 13<br>− 9 |
| 1<br>− 1 | 11<br>− 9 | 10<br>− 4 | 9<br>− 2 | 14<br>− 6 | 8<br>− 0 | 9<br>− 4 | 10<br>− 2 | 6<br>− 5 | 8<br>− 3 |
| 7<br>− 7 | 14<br>− 8 | 12<br>− 9 | 9<br>− 8 | 12<br>− 7 | 9<br>− 9 | 15<br>− 7 | 8<br>− 8 | 14<br>− 9 | 9<br>− 7 |

## A | 64 Addition Facts
*For use with Test 1*

Name _____

Time _____

Add.

| | | | | | | | |
|---|---|---|---|---|---|---|---|
| 7<br>+ 2 | 9<br>+ 4 | 2<br>+ 8 | 6<br>+ 5 | 4<br>+ 4 | 3<br>+ 9 | 8<br>+ 4 | 5<br>+ 7 |
| 9<br>+ 7 | 4<br>+ 7 | 7<br>+ 5 | 5<br>+ 4 | 3<br>+ 4 | 6<br>+ 8 | 2<br>+ 5 | 8<br>+ 8 |
| 6<br>+ 3 | 2<br>+ 9 | 7<br>+ 8 | 8<br>+ 3 | 5<br>+ 9 | 3<br>+ 6 | 9<br>+ 9 | 4<br>+ 9 |
| 5<br>+ 8 | 9<br>+ 5 | 4<br>+ 5 | 8<br>+ 6 | 2<br>+ 3 | 6<br>+ 6 | 5<br>+ 2 | 7<br>+ 3 |
| 3<br>+ 8 | 8<br>+ 9 | 2<br>+ 2 | 7<br>+ 6 | 5<br>+ 5 | 6<br>+ 9 | 3<br>+ 7 | 9<br>+ 8 |
| 4<br>+ 2 | 3<br>+ 3 | 6<br>+ 4 | 4<br>+ 8 | 9<br>+ 3 | 2<br>+ 4 | 8<br>+ 5 | 7<br>+ 9 |
| 7<br>+ 4 | 2<br>+ 6 | 5<br>+ 3 | 9<br>+ 6 | 4<br>+ 3 | 6<br>+ 7 | 3<br>+ 2 | 8<br>+ 7 |
| 5<br>+ 6 | 8<br>+ 2 | 3<br>+ 5 | 6<br>+ 2 | 7<br>+ 7 | 4<br>+ 6 | 9<br>+ 2 | 2<br>+ 7 |

**64 Multiplication Facts**
*For use with Lesson 11*

Name _____

Time _____

Multiply.

| | | | | | | | |
|---|---|---|---|---|---|---|---|
| 5<br>× 6 | 4<br>× 3 | 9<br>× 8 | 7<br>× 5 | 2<br>× 9 | 8<br>× 4 | 9<br>× 3 | 6<br>× 9 |
| 9<br>× 4 | 2<br>× 5 | 7<br>× 6 | 4<br>× 8 | 7<br>× 9 | 5<br>× 4 | 3<br>× 2 | 9<br>× 7 |
| 3<br>× 7 | 8<br>× 5 | 6<br>× 2 | 5<br>× 5 | 3<br>× 5 | 2<br>× 4 | 7<br>× 7 | 8<br>× 9 |
| 6<br>× 4 | 2<br>× 8 | 4<br>× 4 | 8<br>× 2 | 3<br>× 9 | 6<br>× 6 | 9<br>× 9 | 5<br>× 3 |
| 4<br>× 6 | 8<br>× 8 | 5<br>× 7 | 6<br>× 3 | 2<br>× 2 | 7<br>× 4 | 3<br>× 8 | 8<br>× 6 |
| 2<br>× 6 | 5<br>× 9 | 3<br>× 3 | 9<br>× 2 | 6<br>× 7 | 4<br>× 5 | 7<br>× 2 | 9<br>× 6 |
| 5<br>× 2 | 7<br>× 8 | 2<br>× 3 | 6<br>× 8 | 4<br>× 7 | 9<br>× 5 | 3<br>× 6 | 8<br>× 7 |
| 3<br>× 4 | 7<br>× 3 | 5<br>× 8 | 4<br>× 2 | 8<br>× 3 | 2<br>× 7 | 6<br>× 5 | 4<br>× 9 |

*Saxon Math 7/6—Homeschool*

# D | 64 Multiplication Facts

*For use with Lesson 12*

Name _____

Time _____

Multiply.

| | | | | | | | |
|---|---|---|---|---|---|---|---|
| 5<br>× 6 | 4<br>× 3 | 9<br>× 8 | 7<br>× 5 | 2<br>× 9 | 8<br>× 4 | 9<br>× 3 | 6<br>× 9 |
| 9<br>× 4 | 2<br>× 5 | 7<br>× 6 | 4<br>× 8 | 7<br>× 9 | 5<br>× 4 | 3<br>× 2 | 9<br>× 7 |
| 3<br>× 7 | 8<br>× 5 | 6<br>× 2 | 5<br>× 5 | 3<br>× 5 | 2<br>× 4 | 7<br>× 7 | 8<br>× 9 |
| 6<br>× 4 | 2<br>× 8 | 4<br>× 4 | 8<br>× 2 | 3<br>× 9 | 6<br>× 6 | 9<br>× 9 | 5<br>× 3 |
| 4<br>× 6 | 8<br>× 8 | 5<br>× 7 | 6<br>× 3 | 2<br>× 2 | 7<br>× 4 | 3<br>× 8 | 8<br>× 6 |
| 2<br>× 6 | 5<br>× 9 | 3<br>× 3 | 9<br>× 2 | 6<br>× 7 | 4<br>× 5 | 7<br>× 2 | 9<br>× 6 |
| 5<br>× 2 | 7<br>× 8 | 2<br>× 3 | 6<br>× 8 | 4<br>× 7 | 9<br>× 5 | 3<br>× 6 | 8<br>× 7 |
| 3<br>× 4 | 7<br>× 3 | 5<br>× 8 | 4<br>× 2 | 8<br>× 3 | 2<br>× 7 | 6<br>× 5 | 4<br>× 9 |

## A  64 Addition Facts
*For use with Lesson 13*

Name _____

Time _____

Add.

| | | | | | | | |
|---|---|---|---|---|---|---|---|
| 7<br>+ 2 | 9<br>+ 4 | 2<br>+ 8 | 6<br>+ 5 | 4<br>+ 4 | 3<br>+ 9 | 8<br>+ 4 | 5<br>+ 7 |
| 9<br>+ 7 | 4<br>+ 7 | 7<br>+ 5 | 5<br>+ 4 | 3<br>+ 4 | 6<br>+ 8 | 2<br>+ 5 | 8<br>+ 8 |
| 6<br>+ 3 | 2<br>+ 9 | 7<br>+ 8 | 8<br>+ 3 | 5<br>+ 9 | 3<br>+ 6 | 9<br>+ 9 | 4<br>+ 9 |
| 5<br>+ 8 | 9<br>+ 5 | 4<br>+ 5 | 8<br>+ 6 | 2<br>+ 3 | 6<br>+ 6 | 5<br>+ 2 | 7<br>+ 3 |
| 3<br>+ 8 | 8<br>+ 9 | 2<br>+ 2 | 7<br>+ 6 | 5<br>+ 5 | 6<br>+ 9 | 3<br>+ 7 | 9<br>+ 8 |
| 4<br>+ 2 | 3<br>+ 3 | 6<br>+ 4 | 4<br>+ 8 | 9<br>+ 3 | 2<br>+ 4 | 8<br>+ 5 | 7<br>+ 9 |
| 7<br>+ 4 | 2<br>+ 6 | 5<br>+ 3 | 9<br>+ 6 | 4<br>+ 3 | 6<br>+ 7 | 3<br>+ 2 | 8<br>+ 7 |
| 5<br>+ 6 | 8<br>+ 2 | 3<br>+ 5 | 6<br>+ 2 | 7<br>+ 7 | 4<br>+ 6 | 9<br>+ 2 | 2<br>+ 7 |

*Saxon Math 7/6—Homeschool*

**64 Multiplication Facts**
*For use with Lesson 14*

Name _____

Time _____

Multiply.

| | | | | | | | |
|---|---|---|---|---|---|---|---|
| 5<br>× 6 | 4<br>× 3 | 9<br>× 8 | 7<br>× 5 | 2<br>× 9 | 8<br>× 4 | 9<br>× 3 | 6<br>× 9 |
| 9<br>× 4 | 2<br>× 5 | 7<br>× 6 | 4<br>× 8 | 7<br>× 9 | 5<br>× 4 | 3<br>× 2 | 9<br>× 7 |
| 3<br>× 7 | 8<br>× 5 | 6<br>× 2 | 5<br>× 5 | 3<br>× 5 | 2<br>× 4 | 7<br>× 7 | 8<br>× 9 |
| 6<br>× 4 | 2<br>× 8 | 4<br>× 4 | 8<br>× 2 | 3<br>× 9 | 6<br>× 6 | 9<br>× 9 | 5<br>× 3 |
| 4<br>× 6 | 8<br>× 8 | 5<br>× 7 | 6<br>× 3 | 2<br>× 2 | 7<br>× 4 | 3<br>× 8 | 8<br>× 6 |
| 2<br>× 6 | 5<br>× 9 | 3<br>× 3 | 9<br>× 2 | 6<br>× 7 | 4<br>× 5 | 7<br>× 2 | 9<br>× 6 |
| 5<br>× 2 | 7<br>× 8 | 2<br>× 3 | 6<br>× 8 | 4<br>× 7 | 9<br>× 5 | 3<br>× 6 | 8<br>× 7 |
| 3<br>× 4 | 7<br>× 3 | 5<br>× 8 | 4<br>× 2 | 8<br>× 3 | 2<br>× 7 | 6<br>× 5 | 4<br>× 9 |

## C

### 100 Subtraction Facts
*For use with Lesson 15*

Name _____

Time _____

Subtract.

| | | | | | | | | | |
|---|---|---|---|---|---|---|---|---|---|
| 16<br>− 9 | 7<br>− 1 | 18<br>− 9 | 11<br>− 3 | 13<br>− 7 | 8<br>− 2 | 11<br>− 5 | 5<br>− 0 | 17<br>− 9 | 6<br>− 1 |
| 10<br>− 9 | 6<br>− 2 | 13<br>− 4 | 4<br>− 0 | 10<br>− 5 | 5<br>− 1 | 10<br>− 3 | 12<br>− 6 | 10<br>− 1 | 6<br>− 4 |
| 7<br>− 2 | 14<br>− 7 | 8<br>− 1 | 11<br>− 6 | 3<br>− 3 | 16<br>− 7 | 5<br>− 2 | 12<br>− 4 | 3<br>− 0 | 11<br>− 7 |
| 17<br>− 8 | 6<br>− 0 | 10<br>− 6 | 4<br>− 1 | 9<br>− 5 | 9<br>− 0 | 5<br>− 4 | 12<br>− 5 | 4<br>− 2 | 9<br>− 3 |
| 12<br>− 3 | 16<br>− 8 | 9<br>− 1 | 15<br>− 6 | 11<br>− 4 | 13<br>− 5 | 1<br>− 0 | 8<br>− 5 | 9<br>− 6 | 11<br>− 2 |
| 7<br>− 0 | 10<br>− 8 | 6<br>− 3 | 14<br>− 5 | 3<br>− 1 | 8<br>− 6 | 4<br>− 4 | 11<br>− 8 | 3<br>− 2 | 15<br>− 9 |
| 13<br>− 8 | 7<br>− 4 | 10<br>− 7 | 0<br>− 0 | 12<br>− 8 | 5<br>− 5 | 4<br>− 3 | 8<br>− 7 | 7<br>− 3 | 7<br>− 6 |
| 5<br>− 3 | 7<br>− 5 | 2<br>− 1 | 6<br>− 6 | 8<br>− 4 | 2<br>− 2 | 13<br>− 6 | 15<br>− 8 | 2<br>− 0 | 13<br>− 9 |
| 1<br>− 1 | 11<br>− 9 | 10<br>− 4 | 9<br>− 2 | 14<br>− 6 | 8<br>− 0 | 9<br>− 4 | 10<br>− 2 | 6<br>− 5 | 8<br>− 3 |
| 7<br>− 7 | 14<br>− 8 | 12<br>− 9 | 9<br>− 8 | 12<br>− 7 | 9<br>− 9 | 15<br>− 7 | 8<br>− 8 | 14<br>− 9 | 9<br>− 7 |

## C

**100 Subtraction Facts**
*For use with Test 2*

Name _____

Time _____

Subtract.

| | | | | | | | | | |
|---|---|---|---|---|---|---|---|---|---|
| 16<br>− 9 | 7<br>− 1 | 18<br>− 9 | 11<br>− 3 | 13<br>− 7 | 8<br>− 2 | 11<br>− 5 | 5<br>− 0 | 17<br>− 9 | 6<br>− 1 |
| 10<br>− 9 | 6<br>− 2 | 13<br>− 4 | 4<br>− 0 | 10<br>− 5 | 5<br>− 1 | 10<br>− 3 | 12<br>− 6 | 10<br>− 1 | 6<br>− 4 |
| 7<br>− 2 | 14<br>− 7 | 8<br>− 1 | 11<br>− 6 | 3<br>− 3 | 16<br>− 7 | 5<br>− 2 | 12<br>− 4 | 3<br>− 0 | 11<br>− 7 |
| 17<br>− 8 | 6<br>− 0 | 10<br>− 6 | 4<br>− 1 | 9<br>− 5 | 9<br>− 0 | 5<br>− 4 | 12<br>− 5 | 4<br>− 2 | 9<br>− 3 |
| 12<br>− 3 | 16<br>− 8 | 9<br>− 1 | 15<br>− 6 | 11<br>− 4 | 13<br>− 5 | 1<br>− 0 | 8<br>− 5 | 9<br>− 6 | 11<br>− 2 |
| 7<br>− 0 | 10<br>− 8 | 6<br>− 3 | 14<br>− 5 | 3<br>− 1 | 8<br>− 6 | 4<br>− 4 | 11<br>− 8 | 3<br>− 2 | 15<br>− 9 |
| 13<br>− 8 | 7<br>− 4 | 10<br>− 7 | 0<br>− 0 | 12<br>− 8 | 5<br>− 5 | 4<br>− 3 | 8<br>− 7 | 7<br>− 3 | 7<br>− 6 |
| 5<br>− 3 | 7<br>− 5 | 2<br>− 1 | 6<br>− 6 | 8<br>− 4 | 2<br>− 2 | 13<br>− 6 | 15<br>− 8 | 2<br>− 0 | 13<br>− 9 |
| 1<br>− 1 | 11<br>− 9 | 10<br>− 4 | 9<br>− 2 | 14<br>− 6 | 8<br>− 0 | 9<br>− 4 | 10<br>− 2 | 6<br>− 5 | 8<br>− 3 |
| 7<br>− 7 | 14<br>− 8 | 12<br>− 9 | 9<br>− 8 | 12<br>− 7 | 9<br>− 9 | 15<br>− 7 | 8<br>− 8 | 14<br>− 9 | 9<br>− 7 |

## D | 64 Multiplication Facts
*For use with Lesson 16*

Name _____

Time _____

Multiply.

| | | | | | | | |
|---|---|---|---|---|---|---|---|
| 5<br>× 6 | 4<br>× 3 | 9<br>× 8 | 7<br>× 5 | 2<br>× 9 | 8<br>× 4 | 9<br>× 3 | 6<br>× 9 |
| 9<br>× 4 | 2<br>× 5 | 7<br>× 6 | 4<br>× 8 | 7<br>× 9 | 5<br>× 4 | 3<br>× 2 | 9<br>× 7 |
| 3<br>× 7 | 8<br>× 5 | 6<br>× 2 | 5<br>× 5 | 3<br>× 5 | 2<br>× 4 | 7<br>× 7 | 8<br>× 9 |
| 6<br>× 4 | 2<br>× 8 | 4<br>× 4 | 8<br>× 2 | 3<br>× 9 | 6<br>× 6 | 9<br>× 9 | 5<br>× 3 |
| 4<br>× 6 | 8<br>× 8 | 5<br>× 7 | 6<br>× 3 | 2<br>× 2 | 7<br>× 4 | 3<br>× 8 | 8<br>× 6 |
| 2<br>× 6 | 5<br>× 9 | 3<br>× 3 | 9<br>× 2 | 6<br>× 7 | 4<br>× 5 | 7<br>× 2 | 9<br>× 6 |
| 5<br>× 2 | 7<br>× 8 | 2<br>× 3 | 6<br>× 8 | 4<br>× 7 | 9<br>× 5 | 3<br>× 6 | 8<br>× 7 |
| 3<br>× 4 | 7<br>× 3 | 5<br>× 8 | 4<br>× 2 | 8<br>× 3 | 2<br>× 7 | 6<br>× 5 | 4<br>× 9 |

*Saxon Math 7/6—Homeschool*

| E |
|---|

**100 Multiplication Facts**
*For use with Lesson 17*

Name _____

Time _____

Multiply.

| | | | | | | | | | |
|---|---|---|---|---|---|---|---|---|---|
| 9<br>× 9 | 3<br>× 5 | 8<br>× 5 | 2<br>× 6 | 4<br>× 7 | 0<br>× 3 | 7<br>× 2 | 1<br>× 5 | 7<br>× 8 | 4<br>× 0 |
| 3<br>× 4 | 5<br>× 9 | 0<br>× 2 | 7<br>× 3 | 4<br>× 1 | 2<br>× 7 | 6<br>× 3 | 5<br>× 4 | 1<br>× 0 | 9<br>× 2 |
| 1<br>× 1 | 9<br>× 0 | 2<br>× 8 | 6<br>× 4 | 0<br>× 7 | 8<br>× 1 | 3<br>× 3 | 4<br>× 8 | 9<br>× 3 | 2<br>× 0 |
| 4<br>× 9 | 7<br>× 0 | 1<br>× 2 | 8<br>× 4 | 6<br>× 5 | 2<br>× 9 | 9<br>× 4 | 0<br>× 1 | 7<br>× 4 | 5<br>× 8 |
| 0<br>× 8 | 4<br>× 2 | 9<br>× 8 | 3<br>× 6 | 5<br>× 5 | 1<br>× 6 | 5<br>× 0 | 6<br>× 6 | 2<br>× 1 | 7<br>× 9 |
| 9<br>× 1 | 2<br>× 2 | 5<br>× 1 | 4<br>× 3 | 0<br>× 0 | 8<br>× 9 | 3<br>× 7 | 9<br>× 7 | 1<br>× 7 | 6<br>× 0 |
| 5<br>× 6 | 7<br>× 5 | 3<br>× 0 | 8<br>× 8 | 1<br>× 3 | 8<br>× 3 | 5<br>× 2 | 0<br>× 4 | 9<br>× 5 | 6<br>× 7 |
| 2<br>× 3 | 8<br>× 6 | 0<br>× 5 | 6<br>× 1 | 3<br>× 8 | 7<br>× 6 | 1<br>× 8 | 9<br>× 6 | 4<br>× 4 | 5<br>× 3 |
| 7<br>× 7 | 1<br>× 4 | 6<br>× 2 | 4<br>× 5 | 2<br>× 4 | 8<br>× 0 | 3<br>× 1 | 6<br>× 8 | 0<br>× 9 | 8<br>× 7 |
| 3<br>× 2 | 4<br>× 6 | 1<br>× 9 | 5<br>× 7 | 8<br>× 2 | 0<br>× 6 | 7<br>× 1 | 2<br>× 5 | 6<br>× 9 | 3<br>× 9 |

# B 100 Addition Facts

For use with Lesson 18

Name _____

Time _____

Add.

| | | | | | | | | | |
|---|---|---|---|---|---|---|---|---|---|
| 3 <br> + 2 | 8 <br> + 3 | 2 <br> + 1 | 5 <br> + 6 | 2 <br> + 9 | 4 <br> + 8 | 8 <br> + 0 | 3 <br> + 9 | 1 <br> + 0 | 6 <br> + 3 |
| 7 <br> + 3 | 1 <br> + 6 | 4 <br> + 7 | 0 <br> + 3 | 6 <br> + 4 | 5 <br> + 5 | 3 <br> + 1 | 7 <br> + 2 | 8 <br> + 5 | 2 <br> + 5 |
| 4 <br> + 0 | 5 <br> + 7 | 1 <br> + 1 | 5 <br> + 4 | 2 <br> + 8 | 7 <br> + 1 | 4 <br> + 6 | 0 <br> + 2 | 6 <br> + 5 | 4 <br> + 9 |
| 8 <br> + 6 | 0 <br> + 4 | 5 <br> + 8 | 7 <br> + 4 | 1 <br> + 7 | 6 <br> + 6 | 4 <br> + 1 | 8 <br> + 2 | 2 <br> + 4 | 6 <br> + 0 |
| 9 <br> + 1 | 8 <br> + 8 | 2 <br> + 2 | 4 <br> + 5 | 6 <br> + 2 | 0 <br> + 0 | 5 <br> + 9 | 3 <br> + 3 | 8 <br> + 1 | 2 <br> + 7 |
| 4 <br> + 4 | 7 <br> + 5 | 0 <br> + 1 | 8 <br> + 7 | 3 <br> + 4 | 7 <br> + 9 | 1 <br> + 2 | 6 <br> + 7 | 0 <br> + 8 | 9 <br> + 2 |
| 0 <br> + 9 | 8 <br> + 9 | 7 <br> + 6 | 1 <br> + 3 | 6 <br> + 8 | 2 <br> + 0 | 8 <br> + 4 | 3 <br> + 5 | 9 <br> + 8 | 5 <br> + 0 |
| 9 <br> + 3 | 2 <br> + 6 | 3 <br> + 0 | 6 <br> + 1 | 3 <br> + 6 | 5 <br> + 2 | 0 <br> + 5 | 6 <br> + 9 | 1 <br> + 8 | 9 <br> + 6 |
| 4 <br> + 3 | 9 <br> + 9 | 0 <br> + 7 | 9 <br> + 4 | 7 <br> + 7 | 1 <br> + 4 | 3 <br> + 7 | 7 <br> + 0 | 2 <br> + 3 | 5 <br> + 1 |
| 9 <br> + 5 | 1 <br> + 5 | 9 <br> + 0 | 3 <br> + 8 | 1 <br> + 9 | 5 <br> + 3 | 4 <br> + 2 | 9 <br> + 7 | 0 <br> + 6 | 7 <br> + 8 |

### Test-Scores Line Graph
*For use with Lesson 18*

Name _____

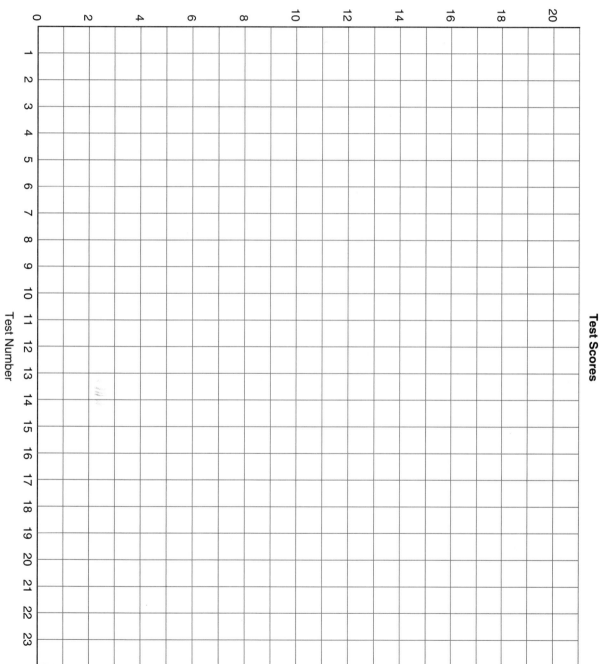

Number of Correct Answers

Test Scores

## D | 64 Multiplication Facts
*For use with Lesson 19*

Name _____

Time _____

Multiply.

| | | | | | | | |
|---|---|---|---|---|---|---|---|
| 5<br>× 6 | 4<br>× 3 | 9<br>× 8 | 7<br>× 5 | 2<br>× 9 | 8<br>× 4 | 9<br>× 3 | 6<br>× 9 |
| 9<br>× 4 | 2<br>× 5 | 7<br>× 6 | 4<br>× 8 | 7<br>× 9 | 5<br>× 4 | 3<br>× 2 | 9<br>× 7 |
| 3<br>× 7 | 8<br>× 5 | 6<br>× 2 | 5<br>× 5 | 3<br>× 5 | 2<br>× 4 | 7<br>× 7 | 8<br>× 9 |
| 6<br>× 4 | 2<br>× 8 | 4<br>× 4 | 8<br>× 2 | 3<br>× 9 | 6<br>× 6 | 9<br>× 9 | 5<br>× 3 |
| 4<br>× 6 | 8<br>× 8 | 5<br>× 7 | 6<br>× 3 | 2<br>× 2 | 7<br>× 4 | 3<br>× 8 | 8<br>× 6 |
| 2<br>× 6 | 5<br>× 9 | 3<br>× 3 | 9<br>× 2 | 6<br>× 7 | 4<br>× 5 | 7<br>× 2 | 9<br>× 6 |
| 5<br>× 2 | 7<br>× 8 | 2<br>× 3 | 6<br>× 8 | 4<br>× 7 | 9<br>× 5 | 3<br>× 6 | 8<br>× 7 |
| 3<br>× 4 | 7<br>× 3 | 5<br>× 8 | 4<br>× 2 | 8<br>× 3 | 2<br>× 7 | 6<br>× 5 | 4<br>× 9 |

## C  100 Subtraction Facts
*For use with Lesson 20*

Name _____

Time _____

Subtract.

| | | | | | | | | | |
|---|---|---|---|---|---|---|---|---|---|
| 16<br>− 9 | 7<br>− 1 | 18<br>− 9 | 11<br>− 3 | 13<br>− 7 | 8<br>− 2 | 11<br>− 5 | 5<br>− 0 | 17<br>− 9 | 6<br>− 1 |
| 10<br>− 9 | 6<br>− 2 | 13<br>− 4 | 4<br>− 0 | 10<br>− 5 | 5<br>− 1 | 10<br>− 3 | 12<br>− 6 | 10<br>− 1 | 6<br>− 4 |
| 7<br>− 2 | 14<br>− 7 | 8<br>− 1 | 11<br>− 6 | 3<br>− 3 | 16<br>− 7 | 5<br>− 2 | 12<br>− 4 | 3<br>− 0 | 11<br>− 7 |
| 17<br>− 8 | 6<br>− 0 | 10<br>− 6 | 4<br>− 1 | 9<br>− 5 | 9<br>− 0 | 5<br>− 4 | 12<br>− 5 | 4<br>− 2 | 9<br>− 3 |
| 12<br>− 3 | 16<br>− 8 | 9<br>− 1 | 15<br>− 6 | 11<br>− 4 | 13<br>− 5 | 1<br>− 0 | 8<br>− 5 | 9<br>− 6 | 11<br>− 2 |
| 7<br>− 0 | 10<br>− 8 | 6<br>− 3 | 14<br>− 5 | 3<br>− 1 | 8<br>− 6 | 4<br>− 4 | 11<br>− 8 | 3<br>− 2 | 15<br>− 9 |
| 13<br>− 8 | 7<br>− 4 | 10<br>− 7 | 0<br>− 0 | 12<br>− 8 | 5<br>− 5 | 4<br>− 3 | 8<br>− 7 | 7<br>− 3 | 7<br>− 6 |
| 5<br>− 3 | 7<br>− 5 | 2<br>− 1 | 6<br>− 6 | 8<br>− 4 | 2<br>− 2 | 13<br>− 6 | 15<br>− 8 | 2<br>− 0 | 13<br>− 9 |
| 1<br>− 1 | 11<br>− 9 | 10<br>− 4 | 9<br>− 2 | 14<br>− 6 | 8<br>− 0 | 9<br>− 4 | 10<br>− 2 | 6<br>− 5 | 8<br>− 3 |
| 7<br>− 7 | 14<br>− 8 | 12<br>− 9 | 9<br>− 8 | 12<br>− 7 | 9<br>− 9 | 15<br>− 7 | 8<br>− 8 | 14<br>− 9 | 9<br>− 7 |

## D | 64 Multiplication Facts
*For use with Test 3*

Name _____

Time _____

Multiply.

| | | | | | | | |
|---|---|---|---|---|---|---|---|
| 5<br>× 6 | 4<br>× 3 | 9<br>× 8 | 7<br>× 5 | 2<br>× 9 | 8<br>× 4 | 9<br>× 3 | 6<br>× 9 |
| 9<br>× 4 | 2<br>× 5 | 7<br>× 6 | 4<br>× 8 | 7<br>× 9 | 5<br>× 4 | 3<br>× 2 | 9<br>× 7 |
| 3<br>× 7 | 8<br>× 5 | 6<br>× 2 | 5<br>× 5 | 3<br>× 5 | 2<br>× 4 | 7<br>× 7 | 8<br>× 9 |
| 6<br>× 4 | 2<br>× 8 | 4<br>× 4 | 8<br>× 2 | 3<br>× 9 | 6<br>× 6 | 9<br>× 9 | 5<br>× 3 |
| 4<br>× 6 | 8<br>× 8 | 5<br>× 7 | 6<br>× 3 | 2<br>× 2 | 7<br>× 4 | 3<br>× 8 | 8<br>× 6 |
| 2<br>× 6 | 5<br>× 9 | 3<br>× 3 | 9<br>× 2 | 6<br>× 7 | 4<br>× 5 | 7<br>× 2 | 9<br>× 6 |
| 5<br>× 2 | 7<br>× 8 | 2<br>× 3 | 6<br>× 8 | 4<br>× 7 | 9<br>× 5 | 3<br>× 6 | 8<br>× 7 |
| 3<br>× 4 | 7<br>× 3 | 5<br>× 8 | 4<br>× 2 | 8<br>× 3 | 2<br>× 7 | 6<br>× 5 | 4<br>× 9 |

*Saxon Math 7/6—Homeschool*

**D** | **64 Multiplication Facts**
For use with *Lesson 21*

Name _____

Time _____

Multiply.

| | | | | | | | |
|---|---|---|---|---|---|---|---|
| 5<br>× 6 | 4<br>× 3 | 9<br>× 8 | 7<br>× 5 | 2<br>× 9 | 8<br>× 4 | 9<br>× 3 | 6<br>× 9 |
| 9<br>× 4 | 2<br>× 5 | 7<br>× 6 | 4<br>× 8 | 7<br>× 9 | 5<br>× 4 | 3<br>× 2 | 9<br>× 7 |
| 3<br>× 7 | 8<br>× 5 | 6<br>× 2 | 5<br>× 5 | 3<br>× 5 | 2<br>× 4 | 7<br>× 7 | 8<br>× 9 |
| 6<br>× 4 | 2<br>× 8 | 4<br>× 4 | 8<br>× 2 | 3<br>× 9 | 6<br>× 6 | 9<br>× 9 | 5<br>× 3 |
| 4<br>× 6 | 8<br>× 8 | 5<br>× 7 | 6<br>× 3 | 2<br>× 2 | 7<br>× 4 | 3<br>× 8 | 8<br>× 6 |
| 2<br>× 6 | 5<br>× 9 | 3<br>× 3 | 9<br>× 2 | 6<br>× 7 | 4<br>× 5 | 7<br>× 2 | 9<br>× 6 |
| 5<br>× 2 | 7<br>× 8 | 2<br>× 3 | 6<br>× 8 | 4<br>× 7 | 9<br>× 5 | 3<br>× 6 | 8<br>× 7 |
| 3<br>× 4 | 7<br>× 3 | 5<br>× 8 | 4<br>× 2 | 8<br>× 3 | 2<br>× 7 | 6<br>× 5 | 4<br>× 9 |

# C

## 100 Subtraction Facts
*For use with Lesson 22*

Name _____

Time _____

Subtract.

| | | | | | | | | | |
|---|---|---|---|---|---|---|---|---|---|
| 16 − 9 | 7 − 1 | 18 − 9 | 11 − 3 | 13 − 7 | 8 − 2 | 11 − 5 | 5 − 0 | 17 − 9 | 6 − 1 |
| 10 − 9 | 6 − 2 | 13 − 4 | 4 − 0 | 10 − 5 | 5 − 1 | 10 − 3 | 12 − 6 | 10 − 1 | 6 − 4 |
| 7 − 2 | 14 − 7 | 8 − 1 | 11 − 6 | 3 − 3 | 16 − 7 | 5 − 2 | 12 − 4 | 3 − 0 | 11 − 7 |
| 17 − 8 | 6 − 0 | 10 − 6 | 4 − 1 | 9 − 5 | 9 − 0 | 5 − 4 | 12 − 5 | 4 − 2 | 9 − 3 |
| 12 − 3 | 16 − 8 | 9 − 1 | 15 − 6 | 11 − 4 | 13 − 5 | 1 − 0 | 8 − 5 | 9 − 6 | 11 − 2 |
| 7 − 0 | 10 − 8 | 6 − 3 | 14 − 5 | 3 − 1 | 8 − 6 | 4 − 4 | 11 − 8 | 3 − 2 | 15 − 9 |
| 13 − 8 | 7 − 4 | 10 − 7 | 0 − 0 | 12 − 8 | 5 − 5 | 4 − 3 | 8 − 7 | 7 − 3 | 7 − 6 |
| 5 − 3 | 7 − 5 | 2 − 1 | 6 − 6 | 8 − 4 | 2 − 2 | 13 − 6 | 15 − 8 | 2 − 0 | 13 − 9 |
| 1 − 1 | 11 − 9 | 10 − 4 | 9 − 2 | 14 − 6 | 8 − 0 | 9 − 4 | 10 − 2 | 6 − 5 | 8 − 3 |
| 7 − 7 | 14 − 8 | 12 − 9 | 9 − 8 | 12 − 7 | 9 − 9 | 15 − 7 | 8 − 8 | 14 − 9 | 9 − 7 |

**D** **64 Multiplication Facts**
*For use with Lesson 23*

Name _____

Time _____

Multiply.

| | | | | | | | |
|---|---|---|---|---|---|---|---|
| 5<br>× 6 | 4<br>× 3 | 9<br>× 8 | 7<br>× 5 | 2<br>× 9 | 8<br>× 4 | 9<br>× 3 | 6<br>× 9 |
| 9<br>× 4 | 2<br>× 5 | 7<br>× 6 | 4<br>× 8 | 7<br>× 9 | 5<br>× 4 | 3<br>× 2 | 9<br>× 7 |
| 3<br>× 7 | 8<br>× 5 | 6<br>× 2 | 5<br>× 5 | 3<br>× 5 | 2<br>× 4 | 7<br>× 7 | 8<br>× 9 |
| 6<br>× 4 | 2<br>× 8 | 4<br>× 4 | 8<br>× 2 | 3<br>× 9 | 6<br>× 6 | 9<br>× 9 | 5<br>× 3 |
| 4<br>× 6 | 8<br>× 8 | 5<br>× 7 | 6<br>× 3 | 2<br>× 2 | 7<br>× 4 | 3<br>× 8 | 8<br>× 6 |
| 2<br>× 6 | 5<br>× 9 | 3<br>× 3 | 9<br>× 2 | 6<br>× 7 | 4<br>× 5 | 7<br>× 2 | 9<br>× 6 |
| 5<br>× 2 | 7<br>× 8 | 2<br>× 3 | 6<br>× 8 | 4<br>× 7 | 9<br>× 5 | 3<br>× 6 | 8<br>× 7 |
| 3<br>× 4 | 7<br>× 3 | 5<br>× 8 | 4<br>× 2 | 8<br>× 3 | 2<br>× 7 | 6<br>× 5 | 4<br>× 9 |

**100 Subtraction Facts**
*For use with Lesson 24*

Name _____

Time _____

Subtract.

| | | | | | | | | | |
|---|---|---|---|---|---|---|---|---|---|
| 16 − 9 | 7 − 1 | 18 − 9 | 11 − 3 | 13 − 7 | 8 − 2 | 11 − 5 | 5 − 0 | 17 − 9 | 6 − 1 |
| 10 − 9 | 6 − 2 | 13 − 4 | 4 − 0 | 10 − 5 | 5 − 1 | 10 − 3 | 12 − 6 | 10 − 1 | 6 − 4 |
| 7 − 2 | 14 − 7 | 8 − 1 | 11 − 6 | 3 − 3 | 16 − 7 | 5 − 2 | 12 − 4 | 3 − 0 | 11 − 7 |
| 17 − 8 | 6 − 0 | 10 − 6 | 4 − 1 | 9 − 5 | 9 − 0 | 5 − 4 | 12 − 5 | 4 − 2 | 9 − 3 |
| 12 − 3 | 16 − 8 | 9 − 1 | 15 − 6 | 11 − 4 | 13 − 5 | 1 − 0 | 8 − 5 | 9 − 6 | 11 − 2 |
| 7 − 0 | 10 − 8 | 6 − 3 | 14 − 5 | 3 − 1 | 8 − 6 | 4 − 4 | 11 − 8 | 3 − 2 | 15 − 9 |
| 13 − 8 | 7 − 4 | 10 − 7 | 0 − 0 | 12 − 8 | 5 − 5 | 4 − 3 | 8 − 7 | 7 − 3 | 7 − 6 |
| 5 − 3 | 7 − 5 | 2 − 1 | 6 − 6 | 8 − 4 | 2 − 2 | 13 − 6 | 15 − 8 | 2 − 0 | 13 − 9 |
| 1 − 1 | 11 − 9 | 10 − 4 | 9 − 2 | 14 − 6 | 8 − 0 | 9 − 4 | 10 − 2 | 6 − 5 | 8 − 3 |
| 7 − 7 | 14 − 8 | 12 − 9 | 9 − 8 | 12 − 7 | 9 − 9 | 15 − 7 | 8 − 8 | 14 − 9 | 9 − 7 |

*Saxon Math 7/6—Homeschool*

| F |
|---|

**90 Division Facts**
*For use with Lesson 25*

Name _____

Time _____

Divide.

| | | | | | | | | | |
|---|---|---|---|---|---|---|---|---|---|
| 7)21 | 2)10 | 6)42 | 1)3 | 4)24 | 3)6 | 9)54 | 6)18 | 4)0 | 5)30 |
| 4)32 | 8)56 | 1)0 | 6)12 | 3)18 | 9)72 | 5)15 | 2)8 | 7)42 | 6)36 |
| 6)0 | 5)10 | 9)9 | 2)6 | 7)63 | 4)16 | 8)48 | 1)2 | 5)35 | 3)21 |
| 2)18 | 6)6 | 3)15 | 8)40 | 2)0 | 5)20 | 9)27 | 1)8 | 4)4 | 7)35 |
| 4)20 | 9)63 | 1)4 | 7)14 | 3)3 | 8)24 | 5)0 | 6)24 | 8)8 | 2)16 |
| 5)5 | 8)64 | 3)0 | 4)28 | 7)49 | 2)4 | 9)81 | 3)12 | 6)30 | 1)5 |
| 8)32 | 1)1 | 9)36 | 3)27 | 2)14 | 5)25 | 6)48 | 8)0 | 7)28 | 4)36 |
| 2)12 | 5)45 | 1)7 | 4)8 | 7)0 | 8)16 | 3)24 | 9)45 | 1)9 | 6)54 |
| 7)56 | 9)0 | 8)72 | 2)2 | 5)40 | 3)9 | 9)18 | 1)6 | 4)12 | 7)7 |

## D  **64 Multiplication Facts**
*For use with Test 4*

Name _____

Time _____

Multiply.

| | | | | | | | |
|---|---|---|---|---|---|---|---|
| 5<br>× 6 | 4<br>× 3 | 9<br>× 8 | 7<br>× 5 | 2<br>× 9 | 8<br>× 4 | 9<br>× 3 | 6<br>× 9 |
| 9<br>× 4 | 2<br>× 5 | 7<br>× 6 | 4<br>× 8 | 7<br>× 9 | 5<br>× 4 | 3<br>× 2 | 9<br>× 7 |
| 3<br>× 7 | 8<br>× 5 | 6<br>× 2 | 5<br>× 5 | 3<br>× 5 | 2<br>× 4 | 7<br>× 7 | 8<br>× 9 |
| 6<br>× 4 | 2<br>× 8 | 4<br>× 4 | 8<br>× 2 | 3<br>× 9 | 6<br>× 6 | 9<br>× 9 | 5<br>× 3 |
| 4<br>× 6 | 8<br>× 8 | 5<br>× 7 | 6<br>× 3 | 2<br>× 2 | 7<br>× 4 | 3<br>× 8 | 8<br>× 6 |
| 2<br>× 6 | 5<br>× 9 | 3<br>× 3 | 9<br>× 2 | 6<br>× 7 | 4<br>× 5 | 7<br>× 2 | 9<br>× 6 |
| 5<br>× 2 | 7<br>× 8 | 2<br>× 3 | 6<br>× 8 | 4<br>× 7 | 9<br>× 5 | 3<br>× 6 | 8<br>× 7 |
| 3<br>× 4 | 7<br>× 3 | 5<br>× 8 | 4<br>× 2 | 8<br>× 3 | 2<br>× 7 | 6<br>× 5 | 4<br>× 9 |

## 100 Subtraction Facts

*For use with Lesson 26*

Name _____

Time _____

Subtract.

| | | | | | | | | | |
|---|---|---|---|---|---|---|---|---|---|
| 16<br>− 9 | 7<br>− 1 | 18<br>− 9 | 11<br>− 3 | 13<br>− 7 | 8<br>− 2 | 11<br>− 5 | 5<br>− 0 | 17<br>− 9 | 6<br>− 1 |
| 10<br>− 9 | 6<br>− 2 | 13<br>− 4 | 4<br>− 0 | 10<br>− 5 | 5<br>− 1 | 10<br>− 3 | 12<br>− 6 | 10<br>− 1 | 6<br>− 4 |
| 7<br>− 2 | 14<br>− 7 | 8<br>− 1 | 11<br>− 6 | 3<br>− 3 | 16<br>− 7 | 5<br>− 2 | 12<br>− 4 | 3<br>− 0 | 11<br>− 7 |
| 17<br>− 8 | 6<br>− 0 | 10<br>− 6 | 4<br>− 1 | 9<br>− 5 | 9<br>− 0 | 5<br>− 4 | 12<br>− 5 | 4<br>− 2 | 9<br>− 3 |
| 12<br>− 3 | 16<br>− 8 | 9<br>− 1 | 15<br>− 6 | 11<br>− 4 | 13<br>− 5 | 1<br>− 0 | 8<br>− 5 | 9<br>− 6 | 11<br>− 2 |
| 7<br>− 0 | 10<br>− 8 | 6<br>− 3 | 14<br>− 5 | 3<br>− 1 | 8<br>− 6 | 4<br>− 4 | 11<br>− 8 | 3<br>− 2 | 15<br>− 9 |
| 13<br>− 8 | 7<br>− 4 | 10<br>− 7 | 0<br>− 0 | 12<br>− 8 | 5<br>− 5 | 4<br>− 3 | 8<br>− 7 | 7<br>− 3 | 7<br>− 6 |
| 5<br>− 3 | 7<br>− 5 | 2<br>− 1 | 6<br>− 6 | 8<br>− 4 | 2<br>− 2 | 13<br>− 6 | 15<br>− 8 | 2<br>− 0 | 13<br>− 9 |
| 1<br>− 1 | 11<br>− 9 | 10<br>− 4 | 9<br>− 2 | 14<br>− 6 | 8<br>− 0 | 9<br>− 4 | 10<br>− 2 | 6<br>− 5 | 8<br>− 3 |
| 7<br>− 7 | 14<br>− 8 | 12<br>− 9 | 9<br>− 8 | 12<br>− 7 | 9<br>− 9 | 15<br>− 7 | 8<br>− 8 | 14<br>− 9 | 9<br>− 7 |

| E |
|---|

## 100 Multiplication Facts
*For use with Lesson 27*

Name _____

Time _____

Multiply.

| | | | | | | | | | |
|---|---|---|---|---|---|---|---|---|---|
| 9<br>× 9 | 3<br>× 5 | 8<br>× 5 | 2<br>× 6 | 4<br>× 7 | 0<br>× 3 | 7<br>× 2 | 1<br>× 5 | 7<br>× 8 | 4<br>× 0 |
| 3<br>× 4 | 5<br>× 9 | 0<br>× 2 | 7<br>× 3 | 4<br>× 1 | 2<br>× 7 | 6<br>× 3 | 5<br>× 4 | 1<br>× 0 | 9<br>× 2 |
| 1<br>× 1 | 9<br>× 0 | 2<br>× 8 | 6<br>× 4 | 0<br>× 7 | 8<br>× 1 | 3<br>× 3 | 4<br>× 8 | 9<br>× 3 | 2<br>× 0 |
| 4<br>× 9 | 7<br>× 0 | 1<br>× 2 | 8<br>× 4 | 6<br>× 5 | 2<br>× 9 | 9<br>× 4 | 0<br>× 1 | 7<br>× 4 | 5<br>× 8 |
| 0<br>× 8 | 4<br>× 2 | 9<br>× 8 | 3<br>× 6 | 5<br>× 5 | 1<br>× 6 | 5<br>× 0 | 6<br>× 6 | 2<br>× 1 | 7<br>× 9 |
| 9<br>× 1 | 2<br>× 2 | 5<br>× 1 | 4<br>× 3 | 0<br>× 0 | 8<br>× 9 | 3<br>× 7 | 9<br>× 7 | 1<br>× 7 | 6<br>× 0 |
| 5<br>× 6 | 7<br>× 5 | 3<br>× 0 | 8<br>× 8 | 1<br>× 3 | 8<br>× 3 | 5<br>× 2 | 0<br>× 4 | 9<br>× 5 | 6<br>× 7 |
| 2<br>× 3 | 8<br>× 6 | 0<br>× 5 | 6<br>× 1 | 3<br>× 8 | 7<br>× 6 | 1<br>× 8 | 9<br>× 6 | 4<br>× 4 | 5<br>× 3 |
| 7<br>× 7 | 1<br>× 4 | 6<br>× 2 | 4<br>× 5 | 2<br>× 4 | 8<br>× 0 | 3<br>× 1 | 6<br>× 8 | 0<br>× 9 | 8<br>× 7 |
| 3<br>× 2 | 4<br>× 6 | 1<br>× 9 | 5<br>× 7 | 8<br>× 2 | 0<br>× 6 | 7<br>× 1 | 2<br>× 5 | 6<br>× 9 | 3<br>× 9 |

*Saxon Math 7/6—Homeschool*

**90 Division Facts**
*For use with Lesson 28*

Name _____

Time _____

Divide.

| | | | | | | | | | |
|---|---|---|---|---|---|---|---|---|---|
| 7)21 | 2)10 | 6)42 | 1)3 | 4)24 | 3)6 | 9)54 | 6)18 | 4)0 | 5)30 |
| 4)32 | 8)56 | 1)0 | 6)12 | 3)18 | 9)72 | 5)15 | 2)8 | 7)42 | 6)36 |
| 6)0 | 5)10 | 9)9 | 2)6 | 7)63 | 4)16 | 8)48 | 1)2 | 5)35 | 3)21 |
| 2)18 | 6)6 | 3)15 | 8)40 | 2)0 | 5)20 | 9)27 | 1)8 | 4)4 | 7)35 |
| 4)20 | 9)63 | 1)4 | 7)14 | 3)3 | 8)24 | 5)0 | 6)24 | 8)8 | 2)16 |
| 5)5 | 8)64 | 3)0 | 4)28 | 7)49 | 2)4 | 9)81 | 3)12 | 6)30 | 1)5 |
| 8)32 | 1)1 | 9)36 | 3)27 | 2)14 | 5)25 | 6)48 | 8)0 | 7)28 | 4)36 |
| 2)12 | 5)45 | 1)7 | 4)8 | 7)0 | 8)16 | 3)24 | 9)45 | 1)9 | 6)54 |
| 7)56 | 9)0 | 8)72 | 2)2 | 5)40 | 3)9 | 9)18 | 1)6 | 4)12 | 7)7 |

**B** **100 Addition Facts**
*For use with Lesson 29*

Name _____

Time _____

Add.

| | | | | | | | | | |
|---|---|---|---|---|---|---|---|---|---|
| 3<br>+ 2 | 8<br>+ 3 | 2<br>+ 1 | 5<br>+ 6 | 2<br>+ 9 | 4<br>+ 8 | 8<br>+ 0 | 3<br>+ 9 | 1<br>+ 0 | 6<br>+ 3 |
| 7<br>+ 3 | 1<br>+ 6 | 4<br>+ 7 | 0<br>+ 3 | 6<br>+ 4 | 5<br>+ 5 | 3<br>+ 1 | 7<br>+ 2 | 8<br>+ 5 | 2<br>+ 5 |
| 4<br>+ 0 | 5<br>+ 7 | 1<br>+ 1 | 5<br>+ 4 | 2<br>+ 8 | 7<br>+ 1 | 4<br>+ 6 | 0<br>+ 2 | 6<br>+ 5 | 4<br>+ 9 |
| 8<br>+ 6 | 0<br>+ 4 | 5<br>+ 8 | 7<br>+ 4 | 1<br>+ 7 | 6<br>+ 6 | 4<br>+ 1 | 8<br>+ 2 | 2<br>+ 4 | 6<br>+ 0 |
| 9<br>+ 1 | 8<br>+ 8 | 2<br>+ 2 | 4<br>+ 5 | 6<br>+ 2 | 0<br>+ 0 | 5<br>+ 9 | 3<br>+ 3 | 8<br>+ 1 | 2<br>+ 7 |
| 4<br>+ 4 | 7<br>+ 5 | 0<br>+ 1 | 8<br>+ 7 | 3<br>+ 4 | 7<br>+ 9 | 1<br>+ 2 | 6<br>+ 7 | 0<br>+ 8 | 9<br>+ 2 |
| 0<br>+ 9 | 8<br>+ 9 | 7<br>+ 6 | 1<br>+ 3 | 6<br>+ 8 | 2<br>+ 0 | 8<br>+ 4 | 3<br>+ 5 | 9<br>+ 8 | 5<br>+ 0 |
| 9<br>+ 3 | 2<br>+ 6 | 3<br>+ 0 | 6<br>+ 1 | 3<br>+ 6 | 5<br>+ 2 | 0<br>+ 5 | 6<br>+ 9 | 1<br>+ 8 | 9<br>+ 6 |
| 4<br>+ 3 | 9<br>+ 9 | 0<br>+ 7 | 9<br>+ 4 | 7<br>+ 7 | 1<br>+ 4 | 3<br>+ 7 | 7<br>+ 0 | 2<br>+ 3 | 5<br>+ 1 |
| 9<br>+ 5 | 1<br>+ 5 | 9<br>+ 0 | 3<br>+ 8 | 1<br>+ 9 | 5<br>+ 3 | 4<br>+ 2 | 9<br>+ 7 | 0<br>+ 6 | 7<br>+ 8 |

**100 Multiplication Facts**
*For use with Lesson 30*

Name _____

Time _____

Multiply.

| | | | | | | | | | |
|---|---|---|---|---|---|---|---|---|---|
| 9<br>× 9 | 3<br>× 5 | 8<br>× 5 | 2<br>× 6 | 4<br>× 7 | 0<br>× 3 | 7<br>× 2 | 1<br>× 5 | 7<br>× 8 | 4<br>× 0 |
| 3<br>× 4 | 5<br>× 9 | 0<br>× 2 | 7<br>× 3 | 4<br>× 1 | 2<br>× 7 | 6<br>× 3 | 5<br>× 4 | 1<br>× 0 | 9<br>× 2 |
| 1<br>× 1 | 9<br>× 0 | 2<br>× 8 | 6<br>× 4 | 0<br>× 7 | 8<br>× 1 | 3<br>× 3 | 4<br>× 8 | 9<br>× 3 | 2<br>× 0 |
| 4<br>× 9 | 7<br>× 0 | 1<br>× 2 | 8<br>× 4 | 6<br>× 5 | 2<br>× 9 | 9<br>× 4 | 0<br>× 1 | 7<br>× 4 | 5<br>× 8 |
| 0<br>× 8 | 4<br>× 2 | 9<br>× 8 | 3<br>× 6 | 5<br>× 5 | 1<br>× 6 | 5<br>× 0 | 6<br>× 6 | 2<br>× 1 | 7<br>× 9 |
| 9<br>× 1 | 2<br>× 2 | 5<br>× 1 | 4<br>× 3 | 0<br>× 0 | 8<br>× 9 | 3<br>× 7 | 9<br>× 7 | 1<br>× 7 | 6<br>× 0 |
| 5<br>× 6 | 7<br>× 5 | 3<br>× 0 | 8<br>× 8 | 1<br>× 3 | 8<br>× 3 | 5<br>× 2 | 0<br>× 4 | 9<br>× 5 | 6<br>× 7 |
| 2<br>× 3 | 8<br>× 6 | 0<br>× 5 | 6<br>× 1 | 3<br>× 8 | 7<br>× 6 | 1<br>× 8 | 9<br>× 6 | 4<br>× 4 | 5<br>× 3 |
| 7<br>× 7 | 1<br>× 4 | 6<br>× 2 | 4<br>× 5 | 2<br>× 4 | 8<br>× 0 | 3<br>× 1 | 6<br>× 8 | 0<br>× 9 | 8<br>× 7 |
| 3<br>× 2 | 4<br>× 6 | 1<br>× 9 | 5<br>× 7 | 8<br>× 2 | 0<br>× 6 | 7<br>× 1 | 2<br>× 5 | 6<br>× 9 | 3<br>× 9 |

| E | **100 Multiplication Facts** | Name _____ |
|---|---|---|
| | *For use with Test 5* | Time _____ |

Multiply.

| | | | | | | | | | |
|---|---|---|---|---|---|---|---|---|---|
| 9<br>× 9 | 3<br>× 5 | 8<br>× 5 | 2<br>× 6 | 4<br>× 7 | 0<br>× 3 | 7<br>× 2 | 1<br>× 5 | 7<br>× 8 | 4<br>× 0 |
| 3<br>× 4 | 5<br>× 9 | 0<br>× 2 | 7<br>× 3 | 4<br>× 1 | 2<br>× 7 | 6<br>× 3 | 5<br>× 4 | 1<br>× 0 | 9<br>× 2 |
| 1<br>× 1 | 9<br>× 0 | 2<br>× 8 | 6<br>× 4 | 0<br>× 7 | 8<br>× 1 | 3<br>× 3 | 4<br>× 8 | 9<br>× 3 | 2<br>× 0 |
| 4<br>× 9 | 7<br>× 0 | 1<br>× 2 | 8<br>× 4 | 6<br>× 5 | 2<br>× 9 | 9<br>× 4 | 0<br>× 1 | 7<br>× 4 | 5<br>× 8 |
| 0<br>× 8 | 4<br>× 2 | 9<br>× 8 | 3<br>× 6 | 5<br>× 5 | 1<br>× 6 | 5<br>× 0 | 6<br>× 6 | 2<br>× 1 | 7<br>× 9 |
| 9<br>× 1 | 2<br>× 2 | 5<br>× 1 | 4<br>× 3 | 0<br>× 0 | 8<br>× 9 | 3<br>× 7 | 9<br>× 7 | 1<br>× 7 | 6<br>× 0 |
| 5<br>× 6 | 7<br>× 5 | 3<br>× 0 | 8<br>× 8 | 1<br>× 3 | 8<br>× 3 | 5<br>× 2 | 0<br>× 4 | 9<br>× 5 | 6<br>× 7 |
| 2<br>× 3 | 8<br>× 6 | 0<br>× 5 | 6<br>× 1 | 3<br>× 8 | 7<br>× 6 | 1<br>× 8 | 9<br>× 6 | 4<br>× 4 | 5<br>× 3 |
| 7<br>× 7 | 1<br>× 4 | 6<br>× 2 | 4<br>× 5 | 2<br>× 4 | 8<br>× 0 | 3<br>× 1 | 6<br>× 8 | 0<br>× 9 | 8<br>× 7 |
| 3<br>× 2 | 4<br>× 6 | 1<br>× 9 | 5<br>× 7 | 8<br>× 2 | 0<br>× 6 | 7<br>× 1 | 2<br>× 5 | 6<br>× 9 | 3<br>× 9 |

# 8 | Measuring Angles
*For use with Investigation 3*

Name _____

Use a protractor to find the following angle measures:

**1.** m∠*AOB* _____

**2.** m∠*AOC* _____

**3.** m∠*EOD* _____

**4.** m∠*AOD* _____

**5.** m∠*EOB* _____

**6.** m∠*DOG* _____

**7.** m∠*EOF* _____

**8.** m∠*EOG* _____

**9.** m∠*DOF* _____

**10.** m∠*EOA* _____

**11.** m∠*DOB* _____

**12.** m∠*COF* _____

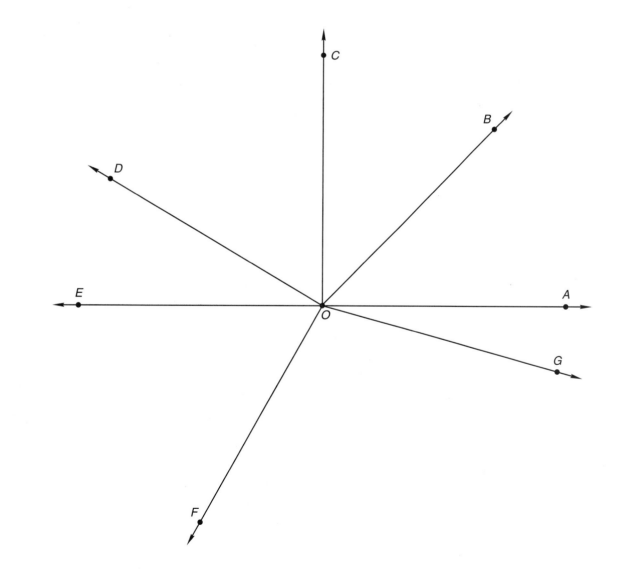

# F

## 90 Division Facts
*For use with Lesson 31*

Name _____

Time _____

Divide.

| | | | | | | | | |
|---|---|---|---|---|---|---|---|---|
| $7\overline{)21}$ | $2\overline{)10}$ | $6\overline{)42}$ | $1\overline{)3}$ | $4\overline{)24}$ | $3\overline{)6}$ | $9\overline{)54}$ | $6\overline{)18}$ | $4\overline{)0}$ | $5\overline{)30}$ |
| $4\overline{)32}$ | $8\overline{)56}$ | $1\overline{)0}$ | $6\overline{)12}$ | $3\overline{)18}$ | $9\overline{)72}$ | $5\overline{)15}$ | $2\overline{)8}$ | $7\overline{)42}$ | $6\overline{)36}$ |
| $6\overline{)0}$ | $5\overline{)10}$ | $9\overline{)9}$ | $2\overline{)6}$ | $7\overline{)63}$ | $4\overline{)16}$ | $8\overline{)48}$ | $1\overline{)2}$ | $5\overline{)35}$ | $3\overline{)21}$ |
| $2\overline{)18}$ | $6\overline{)6}$ | $3\overline{)15}$ | $8\overline{)40}$ | $2\overline{)0}$ | $5\overline{)20}$ | $9\overline{)27}$ | $1\overline{)8}$ | $4\overline{)4}$ | $7\overline{)35}$ |
| $4\overline{)20}$ | $9\overline{)63}$ | $1\overline{)4}$ | $7\overline{)14}$ | $3\overline{)3}$ | $8\overline{)24}$ | $5\overline{)0}$ | $6\overline{)24}$ | $8\overline{)8}$ | $2\overline{)16}$ |
| $5\overline{)5}$ | $8\overline{)64}$ | $3\overline{)0}$ | $4\overline{)28}$ | $7\overline{)49}$ | $2\overline{)4}$ | $9\overline{)81}$ | $3\overline{)12}$ | $6\overline{)30}$ | $1\overline{)5}$ |
| $8\overline{)32}$ | $1\overline{)1}$ | $9\overline{)36}$ | $3\overline{)27}$ | $2\overline{)14}$ | $5\overline{)25}$ | $6\overline{)48}$ | $8\overline{)0}$ | $7\overline{)28}$ | $4\overline{)36}$ |
| $2\overline{)12}$ | $5\overline{)45}$ | $1\overline{)7}$ | $4\overline{)8}$ | $7\overline{)0}$ | $8\overline{)16}$ | $3\overline{)24}$ | $9\overline{)45}$ | $1\overline{)9}$ | $6\overline{)54}$ |
| $7\overline{)56}$ | $9\overline{)0}$ | $8\overline{)72}$ | $2\overline{)2}$ | $5\overline{)40}$ | $3\overline{)9}$ | $9\overline{)18}$ | $1\overline{)6}$ | $4\overline{)12}$ | $7\overline{)7}$ |

## 30 Fractions to Reduce

*For use with Lesson 32*

Name _____

Time _____

Reduce each fraction to lowest terms.

| | | | | |
|---|---|---|---|---|
| $\dfrac{2}{8} =$ | $\dfrac{4}{6} =$ | $\dfrac{6}{10} =$ | $\dfrac{2}{4} =$ | $\dfrac{6}{16} =$ |
| $\dfrac{5}{100} =$ | $\dfrac{9}{12} =$ | $\dfrac{14}{16} =$ | $\dfrac{4}{10} =$ | $\dfrac{4}{12} =$ |
| $\dfrac{2}{10} =$ | $\dfrac{3}{6} =$ | $\dfrac{25}{100} =$ | $\dfrac{3}{12} =$ | $\dfrac{4}{16} =$ |
| $\dfrac{3}{9} =$ | $\dfrac{10}{16} =$ | $\dfrac{6}{9} =$ | $\dfrac{4}{8} =$ | $\dfrac{2}{12} =$ |
| $\dfrac{6}{12} =$ | $\dfrac{2}{16} =$ | $\dfrac{8}{10} =$ | $\dfrac{2}{6} =$ | $\dfrac{75}{100} =$ |
| $\dfrac{12}{16} =$ | $\dfrac{8}{12} =$ | $\dfrac{6}{8} =$ | $\dfrac{10}{12} =$ | $\dfrac{5}{10} =$ |

# G

## 30 Fractions to Reduce

*For use with Lesson 33*

Name _____

Time _____

Reduce each fraction to lowest terms.

| | | | | |
|---|---|---|---|---|
| $\frac{2}{8} =$ | $\frac{4}{6} =$ | $\frac{6}{10} =$ | $\frac{2}{4} =$ | $\frac{6}{16} =$ |
| $\frac{5}{100} =$ | $\frac{9}{12} =$ | $\frac{14}{16} =$ | $\frac{4}{10} =$ | $\frac{4}{12} =$ |
| $\frac{2}{10} =$ | $\frac{3}{6} =$ | $\frac{25}{100} =$ | $\frac{3}{12} =$ | $\frac{4}{16} =$ |
| $\frac{3}{9} =$ | $\frac{10}{16} =$ | $\frac{6}{9} =$ | $\frac{4}{8} =$ | $\frac{2}{12} =$ |
| $\frac{6}{12} =$ | $\frac{2}{16} =$ | $\frac{8}{10} =$ | $\frac{2}{6} =$ | $\frac{75}{100} =$ |
| $\frac{12}{16} =$ | $\frac{8}{12} =$ | $\frac{6}{8} =$ | $\frac{10}{12} =$ | $\frac{5}{10} =$ |

## 64 Multiplication Facts
*For use with Lesson 34*

Name _____

Time _____

Multiply.

| | | | | | | | |
|---|---|---|---|---|---|---|---|
| 5<br>× 6 | 4<br>× 3 | 9<br>× 8 | 7<br>× 5 | 2<br>× 9 | 8<br>× 4 | 9<br>× 3 | 6<br>× 9 |
| 9<br>× 4 | 2<br>× 5 | 7<br>× 6 | 4<br>× 8 | 7<br>× 9 | 5<br>× 4 | 3<br>× 2 | 9<br>× 7 |
| 3<br>× 7 | 8<br>× 5 | 6<br>× 2 | 5<br>× 5 | 3<br>× 5 | 2<br>× 4 | 7<br>× 7 | 8<br>× 9 |
| 6<br>× 4 | 2<br>× 8 | 4<br>× 4 | 8<br>× 2 | 3<br>× 9 | 6<br>× 6 | 9<br>× 9 | 5<br>× 3 |
| 4<br>× 6 | 8<br>× 8 | 5<br>× 7 | 6<br>× 3 | 2<br>× 2 | 7<br>× 4 | 3<br>× 8 | 8<br>× 6 |
| 2<br>× 6 | 5<br>× 9 | 3<br>× 3 | 9<br>× 2 | 6<br>× 7 | 4<br>× 5 | 7<br>× 2 | 9<br>× 6 |
| 5<br>× 2 | 7<br>× 8 | 2<br>× 3 | 6<br>× 8 | 4<br>× 7 | 9<br>× 5 | 3<br>× 6 | 8<br>× 7 |
| 3<br>× 4 | 7<br>× 3 | 5<br>× 8 | 4<br>× 2 | 8<br>× 3 | 2<br>× 7 | 6<br>× 5 | 4<br>× 9 |

**G**    **30 Fractions to Reduce**
*For use with Lesson 35*

Name _____

Time _____

Reduce each fraction to lowest terms.

| | | | | |
|---|---|---|---|---|
| $\frac{2}{8} =$ | $\frac{4}{6} =$ | $\frac{6}{10} =$ | $\frac{2}{4} =$ | $\frac{6}{16} =$ |
| $\frac{5}{100} =$ | $\frac{9}{12} =$ | $\frac{14}{16} =$ | $\frac{4}{10} =$ | $\frac{4}{12} =$ |
| $\frac{2}{10} =$ | $\frac{3}{6} =$ | $\frac{25}{100} =$ | $\frac{3}{12} =$ | $\frac{4}{16} =$ |
| $\frac{3}{9} =$ | $\frac{10}{16} =$ | $\frac{6}{9} =$ | $\frac{4}{8} =$ | $\frac{2}{12} =$ |
| $\frac{6}{12} =$ | $\frac{2}{16} =$ | $\frac{8}{10} =$ | $\frac{2}{6} =$ | $\frac{75}{100} =$ |
| $\frac{12}{16} =$ | $\frac{8}{12} =$ | $\frac{6}{8} =$ | $\frac{10}{12} =$ | $\frac{5}{10} =$ |

| **E** | **100 Multiplication Facts** |
|---|---|

*For use with Test 6*

Name _____

Time _____

Multiply.

| | | | | | | | | | |
|---|---|---|---|---|---|---|---|---|---|
| 9<br>× 9 | 3<br>× 5 | 8<br>× 5 | 2<br>× 6 | 4<br>× 7 | 0<br>× 3 | 7<br>× 2 | 1<br>× 5 | 7<br>× 8 | 4<br>× 0 |
| 3<br>× 4 | 5<br>× 9 | 0<br>× 2 | 7<br>× 3 | 4<br>× 1 | 2<br>× 7 | 6<br>× 3 | 5<br>× 4 | 1<br>× 0 | 9<br>× 2 |
| 1<br>× 1 | 9<br>× 0 | 2<br>× 8 | 6<br>× 4 | 0<br>× 7 | 8<br>× 1 | 3<br>× 3 | 4<br>× 8 | 9<br>× 3 | 2<br>× 0 |
| 4<br>× 9 | 7<br>× 0 | 1<br>× 2 | 8<br>× 4 | 6<br>× 5 | 2<br>× 9 | 9<br>× 4 | 0<br>× 1 | 7<br>× 4 | 5<br>× 8 |
| 0<br>× 8 | 4<br>× 2 | 9<br>× 8 | 3<br>× 6 | 5<br>× 5 | 1<br>× 6 | 5<br>× 0 | 6<br>× 6 | 2<br>× 1 | 7<br>× 9 |
| 9<br>× 1 | 2<br>× 2 | 5<br>× 1 | 4<br>× 3 | 0<br>× 0 | 8<br>× 9 | 3<br>× 7 | 9<br>× 7 | 1<br>× 7 | 6<br>× 0 |
| 5<br>× 6 | 7<br>× 5 | 3<br>× 0 | 8<br>× 8 | 1<br>× 3 | 8<br>× 3 | 5<br>× 2 | 0<br>× 4 | 9<br>× 5 | 6<br>× 7 |
| 2<br>× 3 | 8<br>× 6 | 0<br>× 5 | 6<br>× 1 | 3<br>× 8 | 7<br>× 6 | 1<br>× 8 | 9<br>× 6 | 4<br>× 4 | 5<br>× 3 |
| 7<br>× 7 | 1<br>× 4 | 6<br>× 2 | 4<br>× 5 | 2<br>× 4 | 8<br>× 0 | 3<br>× 1 | 6<br>× 8 | 0<br>× 9 | 8<br>× 7 |
| 3<br>× 2 | 4<br>× 6 | 1<br>× 9 | 5<br>× 7 | 8<br>× 2 | 0<br>× 6 | 7<br>× 1 | 2<br>× 5 | 6<br>× 9 | 3<br>× 9 |

F

**90 Division Facts**
*For use with Lesson 36*

Name _____

Time _____

Divide.

| | | | | | | | | | |
|---|---|---|---|---|---|---|---|---|---|
| $7\overline{)21}$ | $2\overline{)10}$ | $6\overline{)42}$ | $1\overline{)3}$ | $4\overline{)24}$ | $3\overline{)6}$ | $9\overline{)54}$ | $6\overline{)18}$ | $4\overline{)0}$ | $5\overline{)30}$ |
| $4\overline{)32}$ | $8\overline{)56}$ | $1\overline{)0}$ | $6\overline{)12}$ | $3\overline{)18}$ | $9\overline{)72}$ | $5\overline{)15}$ | $2\overline{)8}$ | $7\overline{)42}$ | $6\overline{)36}$ |
| $6\overline{)0}$ | $5\overline{)10}$ | $9\overline{)9}$ | $2\overline{)6}$ | $7\overline{)63}$ | $4\overline{)16}$ | $8\overline{)48}$ | $1\overline{)2}$ | $5\overline{)35}$ | $3\overline{)21}$ |
| $2\overline{)18}$ | $6\overline{)6}$ | $3\overline{)15}$ | $8\overline{)40}$ | $2\overline{)0}$ | $5\overline{)20}$ | $9\overline{)27}$ | $1\overline{)8}$ | $4\overline{)4}$ | $7\overline{)35}$ |
| $4\overline{)20}$ | $9\overline{)63}$ | $1\overline{)4}$ | $7\overline{)14}$ | $3\overline{)3}$ | $8\overline{)24}$ | $5\overline{)0}$ | $6\overline{)24}$ | $8\overline{)8}$ | $2\overline{)16}$ |
| $5\overline{)5}$ | $8\overline{)64}$ | $3\overline{)0}$ | $4\overline{)28}$ | $7\overline{)49}$ | $2\overline{)4}$ | $9\overline{)81}$ | $3\overline{)12}$ | $6\overline{)30}$ | $1\overline{)5}$ |
| $8\overline{)32}$ | $1\overline{)1}$ | $9\overline{)36}$ | $3\overline{)27}$ | $2\overline{)14}$ | $5\overline{)25}$ | $6\overline{)48}$ | $8\overline{)0}$ | $7\overline{)28}$ | $4\overline{)36}$ |
| $2\overline{)12}$ | $5\overline{)45}$ | $1\overline{)7}$ | $4\overline{)8}$ | $7\overline{)0}$ | $8\overline{)16}$ | $3\overline{)24}$ | $9\overline{)45}$ | $1\overline{)9}$ | $6\overline{)54}$ |
| $7\overline{)56}$ | $9\overline{)0}$ | $8\overline{)72}$ | $2\overline{)2}$ | $5\overline{)40}$ | $3\overline{)9}$ | $9\overline{)18}$ | $1\overline{)6}$ | $4\overline{)12}$ | $7\overline{)7}$ |

*Saxon Math 7/6—Homeschool*

## G | 30 Fractions to Reduce
*For use with Lesson 37*

Name _____

Time _____

Reduce each fraction to lowest terms.

| | | | | |
|---|---|---|---|---|
| $\frac{2}{8} =$ | $\frac{4}{6} =$ | $\frac{6}{10} =$ | $\frac{2}{4} =$ | $\frac{6}{16} =$ |
| $\frac{5}{100} =$ | $\frac{9}{12} =$ | $\frac{14}{16} =$ | $\frac{4}{10} =$ | $\frac{4}{12} =$ |
| $\frac{2}{10} =$ | $\frac{3}{6} =$ | $\frac{25}{100} =$ | $\frac{3}{12} =$ | $\frac{4}{16} =$ |
| $\frac{3}{9} =$ | $\frac{10}{16} =$ | $\frac{6}{9} =$ | $\frac{4}{8} =$ | $\frac{2}{12} =$ |
| $\frac{6}{12} =$ | $\frac{2}{16} =$ | $\frac{8}{10} =$ | $\frac{2}{6} =$ | $\frac{75}{100} =$ |
| $\frac{12}{16} =$ | $\frac{8}{12} =$ | $\frac{6}{8} =$ | $\frac{10}{12} =$ | $\frac{5}{10} =$ |

*Saxon Math 7/6—Homeschool*

## A | 64 Addition Facts

*For use with Lesson 38*

Name _____

Time _____

Add.

| | | | | | | | |
|---|---|---|---|---|---|---|---|
| 7<br>+ 2 | 9<br>+ 4 | 2<br>+ 8 | 6<br>+ 5 | 4<br>+ 4 | 3<br>+ 9 | 8<br>+ 4 | 5<br>+ 7 |
| 9<br>+ 7 | 4<br>+ 7 | 7<br>+ 5 | 5<br>+ 4 | 3<br>+ 4 | 6<br>+ 8 | 2<br>+ 5 | 8<br>+ 8 |
| 6<br>+ 3 | 2<br>+ 9 | 7<br>+ 8 | 8<br>+ 3 | 5<br>+ 9 | 3<br>+ 6 | 9<br>+ 9 | 4<br>+ 9 |
| 5<br>+ 8 | 9<br>+ 5 | 4<br>+ 5 | 8<br>+ 6 | 2<br>+ 3 | 6<br>+ 6 | 5<br>+ 2 | 7<br>+ 3 |
| 3<br>+ 8 | 8<br>+ 9 | 2<br>+ 2 | 7<br>+ 6 | 5<br>+ 5 | 6<br>+ 9 | 3<br>+ 7 | 9<br>+ 8 |
| 4<br>+ 2 | 3<br>+ 3 | 6<br>+ 4 | 4<br>+ 8 | 9<br>+ 3 | 2<br>+ 4 | 8<br>+ 5 | 7<br>+ 9 |
| 7<br>+ 4 | 2<br>+ 6 | 5<br>+ 3 | 9<br>+ 6 | 4<br>+ 3 | 6<br>+ 7 | 3<br>+ 2 | 8<br>+ 7 |
| 5<br>+ 6 | 8<br>+ 2 | 3<br>+ 5 | 6<br>+ 2 | 7<br>+ 7 | 4<br>+ 6 | 9<br>+ 2 | 2<br>+ 7 |

**G**  **30 Fractions to Reduce**
*For use with Lesson 39*

Name _____

Time _____

Reduce each fraction to lowest terms.

| | | | | |
|---|---|---|---|---|
| $\frac{2}{8}$ = | $\frac{4}{6}$ = | $\frac{6}{10}$ = | $\frac{2}{4}$ = | $\frac{6}{16}$ = |
| $\frac{5}{100}$ = | $\frac{9}{12}$ = | $\frac{14}{16}$ = | $\frac{4}{10}$ = | $\frac{4}{12}$ = |
| $\frac{2}{10}$ = | $\frac{3}{6}$ = | $\frac{25}{100}$ = | $\frac{3}{12}$ = | $\frac{4}{16}$ = |
| $\frac{3}{9}$ = | $\frac{10}{16}$ = | $\frac{6}{9}$ = | $\frac{4}{8}$ = | $\frac{2}{12}$ = |
| $\frac{6}{12}$ = | $\frac{2}{16}$ = | $\frac{8}{10}$ = | $\frac{2}{6}$ = | $\frac{75}{100}$ = |
| $\frac{12}{16}$ = | $\frac{8}{12}$ = | $\frac{6}{8}$ = | $\frac{10}{12}$ = | $\frac{5}{10}$ = |

| D |
|---|

**64 Multiplication Facts**
*For use with Lesson 40*

Name _____

Time _____

Multiply.

| | | | | | | | |
|---|---|---|---|---|---|---|---|
| 5<br>× 6 | 4<br>× 3 | 9<br>× 8 | 7<br>× 5 | 2<br>× 9 | 8<br>× 4 | 9<br>× 3 | 6<br>× 9 |
| 9<br>× 4 | 2<br>× 5 | 7<br>× 6 | 4<br>× 8 | 7<br>× 9 | 5<br>× 4 | 3<br>× 2 | 9<br>× 7 |
| 3<br>× 7 | 8<br>× 5 | 6<br>× 2 | 5<br>× 5 | 3<br>× 5 | 2<br>× 4 | 7<br>× 7 | 8<br>× 9 |
| 6<br>× 4 | 2<br>× 8 | 4<br>× 4 | 8<br>× 2 | 3<br>× 9 | 6<br>× 6 | 9<br>× 9 | 5<br>× 3 |
| 4<br>× 6 | 8<br>× 8 | 5<br>× 7 | 6<br>× 3 | 2<br>× 2 | 7<br>× 4 | 3<br>× 8 | 8<br>× 6 |
| 2<br>× 6 | 5<br>× 9 | 3<br>× 3 | 9<br>× 2 | 6<br>× 7 | 4<br>× 5 | 7<br>× 2 | 9<br>× 6 |
| 5<br>× 2 | 7<br>× 8 | 2<br>× 3 | 6<br>× 8 | 4<br>× 7 | 9<br>× 5 | 3<br>× 6 | 8<br>× 7 |
| 3<br>× 4 | 7<br>× 3 | 5<br>× 8 | 4<br>× 2 | 8<br>× 3 | 2<br>× 7 | 6<br>× 5 | 4<br>× 9 |

# F

## 90 Division Facts

*For use with Test 7*

Name _____

Time _____

Divide.

| | | | | | | | | | |
|---|---|---|---|---|---|---|---|---|---|
| 7)21 | 2)10 | 6)42 | 1)3 | 4)24 | 3)6 | 9)54 | 6)18 | 4)0 | 5)30 |
| 4)32 | 8)56 | 1)0 | 6)12 | 3)18 | 9)72 | 5)15 | 2)8 | 7)42 | 6)36 |
| 6)0 | 5)10 | 9)9 | 2)6 | 7)63 | 4)16 | 8)48 | 1)2 | 5)35 | 3)21 |
| 2)18 | 6)6 | 3)15 | 8)40 | 2)0 | 5)20 | 9)27 | 1)8 | 4)4 | 7)35 |
| 4)20 | 9)63 | 1)4 | 7)14 | 3)3 | 8)24 | 5)0 | 6)24 | 8)8 | 2)16 |
| 5)5 | 8)64 | 3)0 | 4)28 | 7)49 | 2)4 | 9)81 | 3)12 | 6)30 | 1)5 |
| 8)32 | 1)1 | 9)36 | 3)27 | 2)14 | 5)25 | 6)48 | 8)0 | 7)28 | 4)36 |
| 2)12 | 5)45 | 1)7 | 4)8 | 7)0 | 8)16 | 3)24 | 9)45 | 1)9 | 6)54 |
| 7)56 | 9)0 | 8)72 | 2)2 | 5)40 | 3)9 | 9)18 | 1)6 | 4)12 | 7)7 |

*Saxon Math 7/6—Homeschool*

## G  30 Fractions to Reduce

*For use with Lesson 41*

Name _____

Time _____

Reduce each fraction to lowest terms.

| | | | | |
|---|---|---|---|---|
| $\dfrac{2}{8} =$ | $\dfrac{4}{6} =$ | $\dfrac{6}{10} =$ | $\dfrac{2}{4} =$ | $\dfrac{6}{16} =$ |
| $\dfrac{5}{100} =$ | $\dfrac{9}{12} =$ | $\dfrac{14}{16} =$ | $\dfrac{4}{10} =$ | $\dfrac{4}{12} =$ |
| $\dfrac{2}{10} =$ | $\dfrac{3}{6} =$ | $\dfrac{25}{100} =$ | $\dfrac{3}{12} =$ | $\dfrac{4}{16} =$ |
| $\dfrac{3}{9} =$ | $\dfrac{10}{16} =$ | $\dfrac{6}{9} =$ | $\dfrac{4}{8} =$ | $\dfrac{2}{12} =$ |
| $\dfrac{6}{12} =$ | $\dfrac{2}{16} =$ | $\dfrac{8}{10} =$ | $\dfrac{2}{6} =$ | $\dfrac{75}{100} =$ |
| $\dfrac{12}{16} =$ | $\dfrac{8}{12} =$ | $\dfrac{6}{8} =$ | $\dfrac{10}{12} =$ | $\dfrac{5}{10} =$ |

**G** | **30 Fractions to Reduce**
*For use with Lesson 42*

Name _____

Time _____

Reduce each fraction to lowest terms.

| | | | | |
|---|---|---|---|---|
| $\frac{2}{8} =$ | $\frac{4}{6} =$ | $\frac{6}{10} =$ | $\frac{2}{4} =$ | $\frac{6}{16} =$ |
| $\frac{5}{100} =$ | $\frac{9}{12} =$ | $\frac{14}{16} =$ | $\frac{4}{10} =$ | $\frac{4}{12} =$ |
| $\frac{2}{10} =$ | $\frac{3}{6} =$ | $\frac{25}{100} =$ | $\frac{3}{12} =$ | $\frac{4}{16} =$ |
| $\frac{3}{9} =$ | $\frac{10}{16} =$ | $\frac{6}{9} =$ | $\frac{4}{8} =$ | $\frac{2}{12} =$ |
| $\frac{6}{12} =$ | $\frac{2}{16} =$ | $\frac{8}{10} =$ | $\frac{2}{6} =$ | $\frac{75}{100} =$ |
| $\frac{12}{16} =$ | $\frac{8}{12} =$ | $\frac{6}{8} =$ | $\frac{10}{12} =$ | $\frac{5}{10} =$ |

*Saxon Math 7/6—Homeschool*

## 100 Subtraction Facts

*For use with Lesson 43*

Name _____

Time _____

Subtract.

| | | | | | | | | | |
|---|---|---|---|---|---|---|---|---|---|
| 16<br>− 9 | 7<br>− 1 | 18<br>− 9 | 11<br>− 3 | 13<br>− 7 | 8<br>− 2 | 11<br>− 5 | 5<br>− 0 | 17<br>− 9 | 6<br>− 1 |
| 10<br>− 9 | 6<br>− 2 | 13<br>− 4 | 4<br>− 0 | 10<br>− 5 | 5<br>− 1 | 10<br>− 3 | 12<br>− 6 | 10<br>− 1 | 6<br>− 4 |
| 7<br>− 2 | 14<br>− 7 | 8<br>− 1 | 11<br>− 6 | 3<br>− 3 | 16<br>− 7 | 5<br>− 2 | 12<br>− 4 | 3<br>− 0 | 11<br>− 7 |
| 17<br>− 8 | 6<br>− 0 | 10<br>− 6 | 4<br>− 1 | 9<br>− 5 | 9<br>− 0 | 5<br>− 4 | 12<br>− 5 | 4<br>− 2 | 9<br>− 3 |
| 12<br>− 3 | 16<br>− 8 | 9<br>− 1 | 15<br>− 6 | 11<br>− 4 | 13<br>− 5 | 1<br>− 0 | 8<br>− 5 | 9<br>− 6 | 11<br>− 2 |
| 7<br>− 0 | 10<br>− 8 | 6<br>− 3 | 14<br>− 5 | 3<br>− 1 | 8<br>− 6 | 4<br>− 4 | 11<br>− 8 | 3<br>− 2 | 15<br>− 9 |
| 13<br>− 8 | 7<br>− 4 | 10<br>− 7 | 0<br>− 0 | 12<br>− 8 | 5<br>− 5 | 4<br>− 3 | 8<br>− 7 | 7<br>− 3 | 7<br>− 6 |
| 5<br>− 3 | 7<br>− 5 | 2<br>− 1 | 6<br>− 6 | 8<br>− 4 | 2<br>− 2 | 13<br>− 6 | 15<br>− 8 | 2<br>− 0 | 13<br>− 9 |
| 1<br>− 1 | 11<br>− 9 | 10<br>− 4 | 9<br>− 2 | 14<br>− 6 | 8<br>− 0 | 9<br>− 4 | 10<br>− 2 | 6<br>− 5 | 8<br>− 3 |
| 7<br>− 7 | 14<br>− 8 | 12<br>− 9 | 9<br>− 8 | 12<br>− 7 | 9<br>− 9 | 15<br>− 7 | 8<br>− 8 | 14<br>− 9 | 9<br>− 7 |

**G**

## 30 Fractions to Reduce
*For use with Lesson 44*

Name _____

Time _____

Reduce each fraction to lowest terms.

| | | | | |
|---|---|---|---|---|
| $\frac{2}{8} =$ | $\frac{4}{6} =$ | $\frac{6}{10} =$ | $\frac{2}{4} =$ | $\frac{6}{16} =$ |
| $\frac{5}{100} =$ | $\frac{9}{12} =$ | $\frac{14}{16} =$ | $\frac{4}{10} =$ | $\frac{4}{12} =$ |
| $\frac{2}{10} =$ | $\frac{3}{6} =$ | $\frac{25}{100} =$ | $\frac{3}{12} =$ | $\frac{4}{16} =$ |
| $\frac{3}{9} =$ | $\frac{10}{16} =$ | $\frac{6}{9} =$ | $\frac{4}{8} =$ | $\frac{2}{12} =$ |
| $\frac{6}{12} =$ | $\frac{2}{16} =$ | $\frac{8}{10} =$ | $\frac{2}{6} =$ | $\frac{75}{100} =$ |
| $\frac{12}{16} =$ | $\frac{8}{12} =$ | $\frac{6}{8} =$ | $\frac{10}{12} =$ | $\frac{5}{10} =$ |

*Saxon Math 7/6—Homeschool*

## F | 90 Division Facts
*For use with Lesson 45*

Name _____

Time _____

Divide.

| | | | | | | | | | |
|---|---|---|---|---|---|---|---|---|---|
| 7)21 | 2)10 | 6)42 | 1)3 | 4)24 | 3)6 | 9)54 | 6)18 | 4)0 | 5)30 |
| 4)32 | 8)56 | 1)0 | 6)12 | 3)18 | 9)72 | 5)15 | 2)8 | 7)42 | 6)36 |
| 6)0 | 5)10 | 9)9 | 2)6 | 7)63 | 4)16 | 8)48 | 1)2 | 5)35 | 3)21 |
| 2)18 | 6)6 | 3)15 | 8)40 | 2)0 | 5)20 | 9)27 | 1)8 | 4)4 | 7)35 |
| 4)20 | 9)63 | 1)4 | 7)14 | 3)3 | 8)24 | 5)0 | 6)24 | 8)8 | 2)16 |
| 5)5 | 8)64 | 3)0 | 4)28 | 7)49 | 2)4 | 9)81 | 3)12 | 6)30 | 1)5 |
| 8)32 | 1)1 | 9)36 | 3)27 | 2)14 | 5)25 | 6)48 | 8)0 | 7)28 | 4)36 |
| 2)12 | 5)45 | 1)7 | 4)8 | 7)0 | 8)16 | 3)24 | 9)45 | 1)9 | 6)54 |
| 7)56 | 9)0 | 8)72 | 2)2 | 5)40 | 3)9 | 9)18 | 1)6 | 4)12 | 7)7 |

**30 Fractions to Reduce**

*For use with Test 8*

Name _____

Time _____

Reduce each fraction to lowest terms.

| | | | | |
|---|---|---|---|---|
| $\dfrac{2}{8} =$ | $\dfrac{4}{6} =$ | $\dfrac{6}{10} =$ | $\dfrac{2}{4} =$ | $\dfrac{6}{16} =$ |
| $\dfrac{5}{100} =$ | $\dfrac{9}{12} =$ | $\dfrac{14}{16} =$ | $\dfrac{4}{10} =$ | $\dfrac{4}{12} =$ |
| $\dfrac{2}{10} =$ | $\dfrac{3}{6} =$ | $\dfrac{25}{100} =$ | $\dfrac{3}{12} =$ | $\dfrac{4}{16} =$ |
| $\dfrac{3}{9} =$ | $\dfrac{10}{16} =$ | $\dfrac{6}{9} =$ | $\dfrac{4}{8} =$ | $\dfrac{2}{12} =$ |
| $\dfrac{6}{12} =$ | $\dfrac{2}{16} =$ | $\dfrac{8}{10} =$ | $\dfrac{2}{6} =$ | $\dfrac{75}{100} =$ |
| $\dfrac{12}{16} =$ | $\dfrac{8}{12} =$ | $\dfrac{6}{8} =$ | $\dfrac{10}{12} =$ | $\dfrac{5}{10} =$ |

*Saxon Math 7/6—Homeschool*

| G |
|---|

## 30 Fractions to Reduce
*For use with Lesson 46*

Name _____

Time _____

Reduce each fraction to lowest terms.

| | | | | |
|---|---|---|---|---|
| $\frac{2}{8} =$ | $\frac{4}{6} =$ | $\frac{6}{10} =$ | $\frac{2}{4} =$ | $\frac{6}{16} =$ |
| $\frac{5}{100} =$ | $\frac{9}{12} =$ | $\frac{14}{16} =$ | $\frac{4}{10} =$ | $\frac{4}{12} =$ |
| $\frac{2}{10} =$ | $\frac{3}{6} =$ | $\frac{25}{100} =$ | $\frac{3}{12} =$ | $\frac{4}{16} =$ |
| $\frac{3}{9} =$ | $\frac{10}{16} =$ | $\frac{6}{9} =$ | $\frac{4}{8} =$ | $\frac{2}{12} =$ |
| $\frac{6}{12} =$ | $\frac{2}{16} =$ | $\frac{8}{10} =$ | $\frac{2}{6} =$ | $\frac{75}{100} =$ |
| $\frac{12}{16} =$ | $\frac{8}{12} =$ | $\frac{6}{8} =$ | $\frac{10}{12} =$ | $\frac{5}{10} =$ |

**E** **100 Multiplication Facts**
*For use with Lesson 47*

Name _____

Time _____

Multiply.

| | | | | | | | | | |
|---|---|---|---|---|---|---|---|---|---|
| 9<br>× 9 | 3<br>× 5 | 8<br>× 5 | 2<br>× 6 | 4<br>× 7 | 0<br>× 3 | 7<br>× 2 | 1<br>× 5 | 7<br>× 8 | 4<br>× 0 |
| 3<br>× 4 | 5<br>× 9 | 0<br>× 2 | 7<br>× 3 | 4<br>× 1 | 2<br>× 7 | 6<br>× 3 | 5<br>× 4 | 1<br>× 0 | 9<br>× 2 |
| 1<br>× 1 | 9<br>× 0 | 2<br>× 8 | 6<br>× 4 | 0<br>× 7 | 8<br>× 1 | 3<br>× 3 | 4<br>× 8 | 9<br>× 3 | 2<br>× 0 |
| 4<br>× 9 | 7<br>× 0 | 1<br>× 2 | 8<br>× 4 | 6<br>× 5 | 2<br>× 9 | 9<br>× 4 | 0<br>× 1 | 7<br>× 4 | 5<br>× 8 |
| 0<br>× 8 | 4<br>× 2 | 9<br>× 8 | 3<br>× 6 | 5<br>× 5 | 1<br>× 6 | 5<br>× 0 | 6<br>× 6 | 2<br>× 1 | 7<br>× 9 |
| 9<br>× 1 | 2<br>× 2 | 5<br>× 1 | 4<br>× 3 | 0<br>× 0 | 8<br>× 9 | 3<br>× 7 | 9<br>× 7 | 1<br>× 7 | 6<br>× 0 |
| 5<br>× 6 | 7<br>× 5 | 3<br>× 0 | 8<br>× 8 | 1<br>× 3 | 8<br>× 3 | 5<br>× 2 | 0<br>× 4 | 9<br>× 5 | 6<br>× 7 |
| 2<br>× 3 | 8<br>× 6 | 0<br>× 5 | 6<br>× 1 | 3<br>× 8 | 7<br>× 6 | 1<br>× 8 | 9<br>× 6 | 4<br>× 4 | 5<br>× 3 |
| 7<br>× 7 | 1<br>× 4 | 6<br>× 2 | 4<br>× 5 | 2<br>× 4 | 8<br>× 0 | 3<br>× 1 | 6<br>× 8 | 0<br>× 9 | 8<br>× 7 |
| 3<br>× 2 | 4<br>× 6 | 1<br>× 9 | 5<br>× 7 | 8<br>× 2 | 0<br>× 6 | 7<br>× 1 | 2<br>× 5 | 6<br>× 9 | 3<br>× 9 |

## 9  Finding the Number of Diameters in a Circumference

Name _____

*For use with Lesson 47*

**Part 1:** Estimates

| Object | Approximate Number of Diameters in the Circumference |
|--------|------------------------------------------------------|
|        |                                                      |
|        |                                                      |
|        |                                                      |
|        |                                                      |

**Part 2:** Measures

| Object | Circumference | Diameter | $\dfrac{\text{Circumference}}{\text{Diameter}}$ |
|--------|---------------|----------|-------------------------------------------------|
|        |               |          |                                                 |
|        |               |          |                                                 |
|        |               |          |                                                 |
|        |               |          |                                                 |

## G  30 Fractions to Reduce
*For use with Lesson 48*

Name _____

Time _____

Reduce each fraction to lowest terms.

| | | | | |
|---|---|---|---|---|
| $\frac{2}{8}=$ | $\frac{4}{6}=$ | $\frac{6}{10}=$ | $\frac{2}{4}=$ | $\frac{6}{16}=$ |
| $\frac{5}{100}=$ | $\frac{9}{12}=$ | $\frac{14}{16}=$ | $\frac{4}{10}=$ | $\frac{4}{12}=$ |
| $\frac{2}{10}=$ | $\frac{3}{6}=$ | $\frac{25}{100}=$ | $\frac{3}{12}=$ | $\frac{4}{16}=$ |
| $\frac{3}{9}=$ | $\frac{10}{16}=$ | $\frac{6}{9}=$ | $\frac{4}{8}=$ | $\frac{2}{12}=$ |
| $\frac{6}{12}=$ | $\frac{2}{16}=$ | $\frac{8}{10}=$ | $\frac{2}{6}=$ | $\frac{75}{100}=$ |
| $\frac{12}{16}=$ | $\frac{8}{12}=$ | $\frac{6}{8}=$ | $\frac{10}{12}=$ | $\frac{5}{10}=$ |

H

**72 Multiplication and
Division Facts**
*For use with Lesson 49*

Name _____

Time _____

Multiply or divide as indicated.

| | | | | | | |
|---|---|---|---|---|---|---|
| $\begin{array}{r} 5 \\ \times\ 9 \\ \hline \end{array}$ | $6\,\overline{)\,36}$ | $\begin{array}{r} 4 \\ \times\ 7 \\ \hline \end{array}$ | $8\,\overline{)\,40}$ | $\begin{array}{r} 10 \\ \times\ 3 \\ \hline \end{array}$ | $5\,\overline{)\,20}$ | $\begin{array}{r} 2 \\ \times\ 7 \\ \hline \end{array}$ | $9\,\overline{)\,27}$ |
| $9\,\overline{)\,81}$ | $\begin{array}{r} 5 \\ \times\ 7 \\ \hline \end{array}$ | $6\,\overline{)\,24}$ | $\begin{array}{r} 2 \\ \times\ 3 \\ \hline \end{array}$ | $7\,\overline{)\,42}$ | $\begin{array}{r} 7 \\ \times\ 9 \\ \hline \end{array}$ | $5\,\overline{)\,10}$ | $\begin{array}{r} 4 \\ \times\ 5 \\ \hline \end{array}$ |
| $\begin{array}{r} 2 \\ \times\ 9 \\ \hline \end{array}$ | $7\,\overline{)\,28}$ | $\begin{array}{r} 6 \\ \times\ 9 \\ \hline \end{array}$ | $4\,\overline{)\,8}$ | $\begin{array}{r} 3 \\ \times\ 3 \\ \hline \end{array}$ | $8\,\overline{)\,24}$ | $\begin{array}{r} 10 \\ \times\ 4 \\ \hline \end{array}$ | $6\,\overline{)\,12}$ |
| $9\,\overline{)\,63}$ | $\begin{array}{r} 2 \\ \times\ 5 \\ \hline \end{array}$ | $8\,\overline{)\,64}$ | $\begin{array}{r} 7 \\ \times\ 6 \\ \hline \end{array}$ | $10\,\overline{)\,100}$ | $\begin{array}{r} 3 \\ \times\ 5 \\ \hline \end{array}$ | $9\,\overline{)\,36}$ | $\begin{array}{r} 10 \\ \times\ 8 \\ \hline \end{array}$ |
| $\begin{array}{r} 4 \\ \times\ 8 \\ \hline \end{array}$ | $5\,\overline{)\,25}$ | $\begin{array}{r} 3 \\ \times\ 8 \\ \hline \end{array}$ | $4\,\overline{)\,16}$ | $\begin{array}{r} 10 \\ \times\ 5 \\ \hline \end{array}$ | $7\,\overline{)\,14}$ | $\begin{array}{r} 2 \\ \times\ 2 \\ \hline \end{array}$ | $8\,\overline{)\,48}$ |
| $9\,\overline{)\,54}$ | $\begin{array}{r} 3 \\ \times\ 6 \\ \hline \end{array}$ | $5\,\overline{)\,15}$ | $\begin{array}{r} 2 \\ \times\ 8 \\ \hline \end{array}$ | $7\,\overline{)\,49}$ | $\begin{array}{r} 4 \\ \times\ 6 \\ \hline \end{array}$ | $6\,\overline{)\,30}$ | $\begin{array}{r} 5 \\ \times\ 8 \\ \hline \end{array}$ |
| $\begin{array}{r} 10 \\ \times\ 6 \\ \hline \end{array}$ | $4\,\overline{)\,12}$ | $\begin{array}{r} 7 \\ \times\ 8 \\ \hline \end{array}$ | $6\,\overline{)\,18}$ | $\begin{array}{r} 3 \\ \times\ 4 \\ \hline \end{array}$ | $7\,\overline{)\,35}$ | $\begin{array}{r} 5 \\ \times\ 6 \\ \hline \end{array}$ | $8\,\overline{)\,32}$ |
| $8\,\overline{)\,56}$ | $\begin{array}{r} 2 \\ \times\ 4 \\ \hline \end{array}$ | $8\,\overline{)\,16}$ | $\begin{array}{r} 6 \\ \times\ 8 \\ \hline \end{array}$ | $7\,\overline{)\,21}$ | $\begin{array}{r} 8 \\ \times\ 9 \\ \hline \end{array}$ | $3\,\overline{)\,6}$ | $\begin{array}{r} 3 \\ \times\ 9 \\ \hline \end{array}$ |
| $\begin{array}{r} 10 \\ \times\ 7 \\ \hline \end{array}$ | $9\,\overline{)\,18}$ | $\begin{array}{r} 3 \\ \times\ 7 \\ \hline \end{array}$ | $9\,\overline{)\,45}$ | $\begin{array}{r} 2 \\ \times\ 6 \\ \hline \end{array}$ | $9\,\overline{)\,72}$ | $\begin{array}{r} 4 \\ \times\ 9 \\ \hline \end{array}$ | $9\,\overline{)\,90}$ |

## G | 30 Fractions to Reduce

*For use with Lesson 50*

Name _____

Time _____

Reduce each fraction to lowest terms.

| | | | | |
|---|---|---|---|---|
| $\frac{2}{8} =$ | $\frac{4}{6} =$ | $\frac{6}{10} =$ | $\frac{2}{4} =$ | $\frac{6}{16} =$ |
| $\frac{5}{100} =$ | $\frac{9}{12} =$ | $\frac{14}{16} =$ | $\frac{4}{10} =$ | $\frac{4}{12} =$ |
| $\frac{2}{10} =$ | $\frac{3}{6} =$ | $\frac{25}{100} =$ | $\frac{3}{12} =$ | $\frac{4}{16} =$ |
| $\frac{3}{9} =$ | $\frac{10}{16} =$ | $\frac{6}{9} =$ | $\frac{4}{8} =$ | $\frac{2}{12} =$ |
| $\frac{6}{12} =$ | $\frac{2}{16} =$ | $\frac{8}{10} =$ | $\frac{2}{6} =$ | $\frac{75}{100} =$ |
| $\frac{12}{16} =$ | $\frac{8}{12} =$ | $\frac{6}{8} =$ | $\frac{10}{12} =$ | $\frac{5}{10} =$ |

| G |
|---|

**30 Fractions to Reduce**
*For use with Test 9*

Name _____

Time _____

Reduce each fraction to lowest terms.

| | | | | |
|---|---|---|---|---|
| $\dfrac{2}{8} =$ | $\dfrac{4}{6} =$ | $\dfrac{6}{10} =$ | $\dfrac{2}{4} =$ | $\dfrac{6}{16} =$ |
| $\dfrac{5}{100} =$ | $\dfrac{9}{12} =$ | $\dfrac{14}{16} =$ | $\dfrac{4}{10} =$ | $\dfrac{4}{12} =$ |
| $\dfrac{2}{10} =$ | $\dfrac{3}{6} =$ | $\dfrac{25}{100} =$ | $\dfrac{3}{12} =$ | $\dfrac{4}{16} =$ |
| $\dfrac{3}{9} =$ | $\dfrac{10}{16} =$ | $\dfrac{6}{9} =$ | $\dfrac{4}{8} =$ | $\dfrac{2}{12} =$ |
| $\dfrac{6}{12} =$ | $\dfrac{2}{16} =$ | $\dfrac{8}{10} =$ | $\dfrac{2}{6} =$ | $\dfrac{75}{100} =$ |
| $\dfrac{12}{16} =$ | $\dfrac{8}{12} =$ | $\dfrac{6}{8} =$ | $\dfrac{10}{12} =$ | $\dfrac{5}{10} =$ |

*Saxon Math 7/6—Homeschool*

# H | 72 Multiplication and Division Facts

*For use with Lesson 51*

Name _____

Time _____

Multiply or divide as indicated.

| | | | | | | |
|---|---|---|---|---|---|---|
| $\begin{array}{r} 5 \\ \times\ 9 \\ \hline \end{array}$ | $6\overline{)36}$ | $\begin{array}{r} 4 \\ \times\ 7 \\ \hline \end{array}$ | $8\overline{)40}$ | $\begin{array}{r} 10 \\ \times\ 3 \\ \hline \end{array}$ | $5\overline{)20}$ | $\begin{array}{r} 2 \\ \times\ 7 \\ \hline \end{array}$ | $9\overline{)27}$ |
| $9\overline{)81}$ | $\begin{array}{r} 5 \\ \times\ 7 \\ \hline \end{array}$ | $6\overline{)24}$ | $\begin{array}{r} 2 \\ \times\ 3 \\ \hline \end{array}$ | $7\overline{)42}$ | $\begin{array}{r} 7 \\ \times\ 9 \\ \hline \end{array}$ | $5\overline{)10}$ | $\begin{array}{r} 4 \\ \times\ 5 \\ \hline \end{array}$ |
| $\begin{array}{r} 2 \\ \times\ 9 \\ \hline \end{array}$ | $7\overline{)28}$ | $\begin{array}{r} 6 \\ \times\ 9 \\ \hline \end{array}$ | $4\overline{)8}$ | $\begin{array}{r} 3 \\ \times\ 3 \\ \hline \end{array}$ | $8\overline{)24}$ | $\begin{array}{r} 10 \\ \times\ 4 \\ \hline \end{array}$ | $6\overline{)12}$ |
| $9\overline{)63}$ | $\begin{array}{r} 2 \\ \times\ 5 \\ \hline \end{array}$ | $8\overline{)64}$ | $\begin{array}{r} 7 \\ \times\ 6 \\ \hline \end{array}$ | $10\overline{)100}$ | $\begin{array}{r} 3 \\ \times\ 5 \\ \hline \end{array}$ | $9\overline{)36}$ | $\begin{array}{r} 10 \\ \times\ 8 \\ \hline \end{array}$ |
| $\begin{array}{r} 4 \\ \times\ 8 \\ \hline \end{array}$ | $5\overline{)25}$ | $\begin{array}{r} 3 \\ \times\ 8 \\ \hline \end{array}$ | $4\overline{)16}$ | $\begin{array}{r} 10 \\ \times\ 5 \\ \hline \end{array}$ | $7\overline{)14}$ | $\begin{array}{r} 2 \\ \times\ 2 \\ \hline \end{array}$ | $8\overline{)48}$ |
| $9\overline{)54}$ | $\begin{array}{r} 3 \\ \times\ 6 \\ \hline \end{array}$ | $5\overline{)15}$ | $\begin{array}{r} 2 \\ \times\ 8 \\ \hline \end{array}$ | $7\overline{)49}$ | $\begin{array}{r} 4 \\ \times\ 6 \\ \hline \end{array}$ | $6\overline{)30}$ | $\begin{array}{r} 5 \\ \times\ 8 \\ \hline \end{array}$ |
| $\begin{array}{r} 10 \\ \times\ 6 \\ \hline \end{array}$ | $4\overline{)12}$ | $\begin{array}{r} 7 \\ \times\ 8 \\ \hline \end{array}$ | $6\overline{)18}$ | $\begin{array}{r} 3 \\ \times\ 4 \\ \hline \end{array}$ | $7\overline{)35}$ | $\begin{array}{r} 5 \\ \times\ 6 \\ \hline \end{array}$ | $8\overline{)32}$ |
| $8\overline{)56}$ | $\begin{array}{r} 2 \\ \times\ 4 \\ \hline \end{array}$ | $8\overline{)16}$ | $\begin{array}{r} 6 \\ \times\ 8 \\ \hline \end{array}$ | $7\overline{)21}$ | $\begin{array}{r} 8 \\ \times\ 9 \\ \hline \end{array}$ | $3\overline{)6}$ | $\begin{array}{r} 3 \\ \times\ 9 \\ \hline \end{array}$ |
| $\begin{array}{r} 10 \\ \times\ 7 \\ \hline \end{array}$ | $9\overline{)18}$ | $\begin{array}{r} 3 \\ \times\ 7 \\ \hline \end{array}$ | $9\overline{)45}$ | $\begin{array}{r} 2 \\ \times\ 6 \\ \hline \end{array}$ | $9\overline{)72}$ | $\begin{array}{r} 4 \\ \times\ 9 \\ \hline \end{array}$ | $9\overline{)90}$ |

### 30 Fractions to Reduce
*For use with Lesson 52*

Name _____

Time _____

Reduce each fraction to lowest terms.

| | | | | |
|---|---|---|---|---|
| $\dfrac{2}{8} =$ | $\dfrac{4}{6} =$ | $\dfrac{6}{10} =$ | $\dfrac{2}{4} =$ | $\dfrac{6}{16} =$ |
| $\dfrac{5}{100} =$ | $\dfrac{9}{12} =$ | $\dfrac{14}{16} =$ | $\dfrac{4}{10} =$ | $\dfrac{4}{12} =$ |
| $\dfrac{2}{10} =$ | $\dfrac{3}{6} =$ | $\dfrac{25}{100} =$ | $\dfrac{3}{12} =$ | $\dfrac{4}{16} =$ |
| $\dfrac{3}{9} =$ | $\dfrac{10}{16} =$ | $\dfrac{6}{9} =$ | $\dfrac{4}{8} =$ | $\dfrac{2}{12} =$ |
| $\dfrac{6}{12} =$ | $\dfrac{2}{16} =$ | $\dfrac{8}{10} =$ | $\dfrac{2}{6} =$ | $\dfrac{75}{100} =$ |
| $\dfrac{12}{16} =$ | $\dfrac{8}{12} =$ | $\dfrac{6}{8} =$ | $\dfrac{10}{12} =$ | $\dfrac{5}{10} =$ |

*Saxon Math 7/6—Homeschool*

## 64 Multiplication Facts
*For use with Lesson 53*

Name _____

Time _____

Multiply.

| | | | | | | | |
|---|---|---|---|---|---|---|---|
| 5<br>× 6 | 4<br>× 3 | 9<br>× 8 | 7<br>× 5 | 2<br>× 9 | 8<br>× 4 | 9<br>× 3 | 6<br>× 9 |
| 9<br>× 4 | 2<br>× 5 | 7<br>× 6 | 4<br>× 8 | 7<br>× 9 | 5<br>× 4 | 3<br>× 2 | 9<br>× 7 |
| 3<br>× 7 | 8<br>× 5 | 6<br>× 2 | 5<br>× 5 | 3<br>× 5 | 2<br>× 4 | 7<br>× 7 | 8<br>× 9 |
| 6<br>× 4 | 2<br>× 8 | 4<br>× 4 | 8<br>× 2 | 3<br>× 9 | 6<br>× 6 | 9<br>× 9 | 5<br>× 3 |
| 4<br>× 6 | 8<br>× 8 | 5<br>× 7 | 6<br>× 3 | 2<br>× 2 | 7<br>× 4 | 3<br>× 8 | 8<br>× 6 |
| 2<br>× 6 | 5<br>× 9 | 3<br>× 3 | 9<br>× 2 | 6<br>× 7 | 4<br>× 5 | 7<br>× 2 | 9<br>× 6 |
| 5<br>× 2 | 7<br>× 8 | 2<br>× 3 | 6<br>× 8 | 4<br>× 7 | 9<br>× 5 | 3<br>× 6 | 8<br>× 7 |
| 3<br>× 4 | 7<br>× 3 | 5<br>× 8 | 4<br>× 2 | 8<br>× 3 | 2<br>× 7 | 6<br>× 5 | 4<br>× 9 |

**G**

## 30 Fractions to Reduce
*For use with Lesson 54*

Name _____

Time _____

Reduce each fraction to lowest terms.

| | | | | |
|---|---|---|---|---|
| $\dfrac{2}{8} =$ | $\dfrac{4}{6} =$ | $\dfrac{6}{10} =$ | $\dfrac{2}{4} =$ | $\dfrac{6}{16} =$ |
| $\dfrac{5}{100} =$ | $\dfrac{9}{12} =$ | $\dfrac{14}{16} =$ | $\dfrac{4}{10} =$ | $\dfrac{4}{12} =$ |
| $\dfrac{2}{10} =$ | $\dfrac{3}{6} =$ | $\dfrac{25}{100} =$ | $\dfrac{3}{12} =$ | $\dfrac{4}{16} =$ |
| $\dfrac{3}{9} =$ | $\dfrac{10}{16} =$ | $\dfrac{6}{9} =$ | $\dfrac{4}{8} =$ | $\dfrac{2}{12} =$ |
| $\dfrac{6}{12} =$ | $\dfrac{2}{16} =$ | $\dfrac{8}{10} =$ | $\dfrac{2}{6} =$ | $\dfrac{75}{100} =$ |
| $\dfrac{12}{16} =$ | $\dfrac{8}{12} =$ | $\dfrac{6}{8} =$ | $\dfrac{10}{12} =$ | $\dfrac{5}{10} =$ |

# I
## 28 Improper Fractions to Simplify
*For use with Lesson 55*

Name _____

Time _____

Write each improper fraction as a mixed number or a whole number.

| | | | |
|---|---|---|---|
| $\dfrac{5}{4} =$ | $\dfrac{16}{12} =$ | $\dfrac{12}{8} =$ | $\dfrac{8}{6} =$ |
| $\dfrac{12}{6} =$ | $\dfrac{12}{10} =$ | $\dfrac{6}{4} =$ | $\dfrac{20}{12} =$ |
| $\dfrac{5}{3} =$ | $\dfrac{10}{8} =$ | $\dfrac{25}{10} =$ | $\dfrac{10}{3} =$ |
| $\dfrac{15}{10} =$ | $\dfrac{3}{2} =$ | $\dfrac{9}{6} =$ | $\dfrac{7}{4} =$ |
| $\dfrac{18}{12} =$ | $\dfrac{8}{3} =$ | $\dfrac{15}{6} =$ | $\dfrac{14}{4} =$ |
| $\dfrac{8}{4} =$ | $\dfrac{10}{6} =$ | $\dfrac{5}{2} =$ | $\dfrac{21}{12} =$ |
| $\dfrac{15}{12} =$ | $\dfrac{10}{4} =$ | $\dfrac{15}{8} =$ | $\dfrac{4}{3} =$ |

# H

## 72 Multiplication and Division Facts

*For use with Test 10*

Name _____

Time _____

Multiply or divide as indicated.

| | | | | | | | |
|---|---|---|---|---|---|---|---|
| $\begin{array}{r} 5 \\ \times\ 9 \\ \hline \end{array}$ | $6\overline{)36}$ | $\begin{array}{r} 4 \\ \times\ 7 \\ \hline \end{array}$ | $8\overline{)40}$ | $\begin{array}{r} 10 \\ \times\ 3 \\ \hline \end{array}$ | $5\overline{)20}$ | $\begin{array}{r} 2 \\ \times\ 7 \\ \hline \end{array}$ | $9\overline{)27}$ |
| $9\overline{)81}$ | $\begin{array}{r} 5 \\ \times\ 7 \\ \hline \end{array}$ | $6\overline{)24}$ | $\begin{array}{r} 2 \\ \times\ 3 \\ \hline \end{array}$ | $7\overline{)42}$ | $\begin{array}{r} 7 \\ \times\ 9 \\ \hline \end{array}$ | $5\overline{)10}$ | $\begin{array}{r} 4 \\ \times\ 5 \\ \hline \end{array}$ |
| $\begin{array}{r} 2 \\ \times\ 9 \\ \hline \end{array}$ | $7\overline{)28}$ | $\begin{array}{r} 6 \\ \times\ 9 \\ \hline \end{array}$ | $4\overline{)8}$ | $\begin{array}{r} 3 \\ \times\ 3 \\ \hline \end{array}$ | $8\overline{)24}$ | $\begin{array}{r} 10 \\ \times\ 4 \\ \hline \end{array}$ | $6\overline{)12}$ |
| $9\overline{)63}$ | $\begin{array}{r} 2 \\ \times\ 5 \\ \hline \end{array}$ | $8\overline{)64}$ | $\begin{array}{r} 7 \\ \times\ 6 \\ \hline \end{array}$ | $10\overline{)100}$ | $\begin{array}{r} 3 \\ \times\ 5 \\ \hline \end{array}$ | $9\overline{)36}$ | $\begin{array}{r} 10 \\ \times\ 8 \\ \hline \end{array}$ |
| $\begin{array}{r} 4 \\ \times\ 8 \\ \hline \end{array}$ | $5\overline{)25}$ | $\begin{array}{r} 3 \\ \times\ 8 \\ \hline \end{array}$ | $4\overline{)16}$ | $\begin{array}{r} 10 \\ \times\ 5 \\ \hline \end{array}$ | $7\overline{)14}$ | $\begin{array}{r} 2 \\ \times\ 2 \\ \hline \end{array}$ | $8\overline{)48}$ |
| $9\overline{)54}$ | $\begin{array}{r} 3 \\ \times\ 6 \\ \hline \end{array}$ | $5\overline{)15}$ | $\begin{array}{r} 2 \\ \times\ 8 \\ \hline \end{array}$ | $7\overline{)49}$ | $\begin{array}{r} 4 \\ \times\ 6 \\ \hline \end{array}$ | $6\overline{)30}$ | $\begin{array}{r} 5 \\ \times\ 8 \\ \hline \end{array}$ |
| $\begin{array}{r} 10 \\ \times\ 6 \\ \hline \end{array}$ | $4\overline{)12}$ | $\begin{array}{r} 7 \\ \times\ 8 \\ \hline \end{array}$ | $6\overline{)18}$ | $\begin{array}{r} 3 \\ \times\ 4 \\ \hline \end{array}$ | $7\overline{)35}$ | $\begin{array}{r} 5 \\ \times\ 6 \\ \hline \end{array}$ | $8\overline{)32}$ |
| $8\overline{)56}$ | $\begin{array}{r} 2 \\ \times\ 4 \\ \hline \end{array}$ | $8\overline{)16}$ | $\begin{array}{r} 6 \\ \times\ 8 \\ \hline \end{array}$ | $7\overline{)21}$ | $\begin{array}{r} 8 \\ \times\ 9 \\ \hline \end{array}$ | $3\overline{)6}$ | $\begin{array}{r} 3 \\ \times\ 9 \\ \hline \end{array}$ |
| $\begin{array}{r} 10 \\ \times\ 7 \\ \hline \end{array}$ | $9\overline{)18}$ | $\begin{array}{r} 3 \\ \times\ 7 \\ \hline \end{array}$ | $9\overline{)45}$ | $\begin{array}{r} 2 \\ \times\ 6 \\ \hline \end{array}$ | $9\overline{)72}$ | $\begin{array}{r} 4 \\ \times\ 9 \\ \hline \end{array}$ | $9\overline{)90}$ |

*Saxon Math 7/6—Homeschool*

## G | 30 Fractions to Reduce

*For use with Lesson 56*

Name _____

Time _____

Reduce each fraction to lowest terms.

| | | | | |
|---|---|---|---|---|
| $\frac{2}{8} =$ | $\frac{4}{6} =$ | $\frac{6}{10} =$ | $\frac{2}{4} =$ | $\frac{6}{16} =$ |
| $\frac{5}{100} =$ | $\frac{9}{12} =$ | $\frac{14}{16} =$ | $\frac{4}{10} =$ | $\frac{4}{12} =$ |
| $\frac{2}{10} =$ | $\frac{3}{6} =$ | $\frac{25}{100} =$ | $\frac{3}{12} =$ | $\frac{4}{16} =$ |
| $\frac{3}{9} =$ | $\frac{10}{16} =$ | $\frac{6}{9} =$ | $\frac{4}{8} =$ | $\frac{2}{12} =$ |
| $\frac{6}{12} =$ | $\frac{2}{16} =$ | $\frac{8}{10} =$ | $\frac{2}{6} =$ | $\frac{75}{100} =$ |
| $\frac{12}{16} =$ | $\frac{8}{12} =$ | $\frac{6}{8} =$ | $\frac{10}{12} =$ | $\frac{5}{10} =$ |

| H |
|---|

## 72 Multiplication and
## Division Facts
*For use with Lesson 57*

Name _____

Time _____

Multiply or divide as indicated.

| | | | | | | |
|---|---|---|---|---|---|---|
| $\begin{array}{r} 5 \\ \times\ 9 \\ \hline \end{array}$ | $6\,\overline{)\,36}$ | $\begin{array}{r} 4 \\ \times\ 7 \\ \hline \end{array}$ | $8\,\overline{)\,40}$ | $\begin{array}{r} 10 \\ \times\ \ 3 \\ \hline \end{array}$ | $5\,\overline{)\,20}$ | $\begin{array}{r} 2 \\ \times\ 7 \\ \hline \end{array}$ | $9\,\overline{)\,27}$ |
| $9\,\overline{)\,81}$ | $\begin{array}{r} 5 \\ \times\ 7 \\ \hline \end{array}$ | $6\,\overline{)\,24}$ | $\begin{array}{r} 2 \\ \times\ 3 \\ \hline \end{array}$ | $7\,\overline{)\,42}$ | $\begin{array}{r} 7 \\ \times\ 9 \\ \hline \end{array}$ | $5\,\overline{)\,10}$ | $\begin{array}{r} 4 \\ \times\ 5 \\ \hline \end{array}$ |
| $\begin{array}{r} 2 \\ \times\ 9 \\ \hline \end{array}$ | $7\,\overline{)\,28}$ | $\begin{array}{r} 6 \\ \times\ 9 \\ \hline \end{array}$ | $4\,\overline{)\,8}$ | $\begin{array}{r} 3 \\ \times\ 3 \\ \hline \end{array}$ | $8\,\overline{)\,24}$ | $\begin{array}{r} 10 \\ \times\ \ 4 \\ \hline \end{array}$ | $6\,\overline{)\,12}$ |
| $9\,\overline{)\,63}$ | $\begin{array}{r} 2 \\ \times\ 5 \\ \hline \end{array}$ | $8\,\overline{)\,64}$ | $\begin{array}{r} 7 \\ \times\ 6 \\ \hline \end{array}$ | $10\,\overline{)\,100}$ | $\begin{array}{r} 3 \\ \times\ 5 \\ \hline \end{array}$ | $9\,\overline{)\,36}$ | $\begin{array}{r} 10 \\ \times\ \ 8 \\ \hline \end{array}$ |
| $\begin{array}{r} 4 \\ \times\ 8 \\ \hline \end{array}$ | $5\,\overline{)\,25}$ | $\begin{array}{r} 3 \\ \times\ 8 \\ \hline \end{array}$ | $4\,\overline{)\,16}$ | $\begin{array}{r} 10 \\ \times\ \ 5 \\ \hline \end{array}$ | $7\,\overline{)\,14}$ | $\begin{array}{r} 2 \\ \times\ 2 \\ \hline \end{array}$ | $8\,\overline{)\,48}$ |
| $9\,\overline{)\,54}$ | $\begin{array}{r} 3 \\ \times\ 6 \\ \hline \end{array}$ | $5\,\overline{)\,15}$ | $\begin{array}{r} 2 \\ \times\ 8 \\ \hline \end{array}$ | $7\,\overline{)\,49}$ | $\begin{array}{r} 4 \\ \times\ 6 \\ \hline \end{array}$ | $6\,\overline{)\,30}$ | $\begin{array}{r} 5 \\ \times\ 8 \\ \hline \end{array}$ |
| $\begin{array}{r} 10 \\ \times\ \ 6 \\ \hline \end{array}$ | $4\,\overline{)\,12}$ | $\begin{array}{r} 7 \\ \times\ 8 \\ \hline \end{array}$ | $6\,\overline{)\,18}$ | $\begin{array}{r} 3 \\ \times\ 4 \\ \hline \end{array}$ | $7\,\overline{)\,35}$ | $\begin{array}{r} 5 \\ \times\ 6 \\ \hline \end{array}$ | $8\,\overline{)\,32}$ |
| $8\,\overline{)\,56}$ | $\begin{array}{r} 2 \\ \times\ 4 \\ \hline \end{array}$ | $8\,\overline{)\,16}$ | $\begin{array}{r} 6 \\ \times\ 8 \\ \hline \end{array}$ | $7\,\overline{)\,21}$ | $\begin{array}{r} 8 \\ \times\ 9 \\ \hline \end{array}$ | $3\,\overline{)\,6}$ | $\begin{array}{r} 3 \\ \times\ 9 \\ \hline \end{array}$ |
| $\begin{array}{r} 10 \\ \times\ \ 7 \\ \hline \end{array}$ | $9\,\overline{)\,18}$ | $\begin{array}{r} 3 \\ \times\ 7 \\ \hline \end{array}$ | $9\,\overline{)\,45}$ | $\begin{array}{r} 2 \\ \times\ 6 \\ \hline \end{array}$ | $9\,\overline{)\,72}$ | $\begin{array}{r} 4 \\ \times\ 9 \\ \hline \end{array}$ | $9\,\overline{)\,90}$ |

## I

## 28 Improper Fractions to Simplify
*For use with Lesson 58*

Name _____

Time _____

Write each improper fraction as a mixed number or a whole number.

| | | | |
|---|---|---|---|
| $\frac{5}{4} =$ | $\frac{16}{12} =$ | $\frac{12}{8} =$ | $\frac{8}{6} =$ |
| $\frac{12}{6} =$ | $\frac{12}{10} =$ | $\frac{6}{4} =$ | $\frac{20}{12} =$ |
| $\frac{5}{3} =$ | $\frac{10}{8} =$ | $\frac{25}{10} =$ | $\frac{10}{3} =$ |
| $\frac{15}{10} =$ | $\frac{3}{2} =$ | $\frac{9}{6} =$ | $\frac{7}{4} =$ |
| $\frac{18}{12} =$ | $\frac{8}{3} =$ | $\frac{15}{6} =$ | $\frac{14}{4} =$ |
| $\frac{8}{4} =$ | $\frac{10}{6} =$ | $\frac{5}{2} =$ | $\frac{21}{12} =$ |
| $\frac{15}{12} =$ | $\frac{10}{4} =$ | $\frac{15}{8} =$ | $\frac{4}{3} =$ |

## 30 Fractions to Reduce
*For use with Lesson 59*

Name _____

Time _____

Reduce each fraction to lowest terms.

| | | | | |
|---|---|---|---|---|
| $\frac{2}{8}=$ | $\frac{4}{6}=$ | $\frac{6}{10}=$ | $\frac{2}{4}=$ | $\frac{6}{16}=$ |
| $\frac{5}{100}=$ | $\frac{9}{12}=$ | $\frac{14}{16}=$ | $\frac{4}{10}=$ | $\frac{4}{12}=$ |
| $\frac{2}{10}=$ | $\frac{3}{6}=$ | $\frac{25}{100}=$ | $\frac{3}{12}=$ | $\frac{4}{16}=$ |
| $\frac{3}{9}=$ | $\frac{10}{16}=$ | $\frac{6}{9}=$ | $\frac{4}{8}=$ | $\frac{2}{12}=$ |
| $\frac{6}{12}=$ | $\frac{2}{16}=$ | $\frac{8}{10}=$ | $\frac{2}{6}=$ | $\frac{75}{100}=$ |
| $\frac{12}{16}=$ | $\frac{8}{12}=$ | $\frac{6}{8}=$ | $\frac{10}{12}=$ | $\frac{5}{10}=$ |

## I  28 Improper Fractions to Simplify

*For use with Lesson 60*

Name _____

Time _____

Write each improper fraction as a mixed number or a whole number.

| | | | |
|---|---|---|---|
| $\dfrac{5}{4} =$ | $\dfrac{16}{12} =$ | $\dfrac{12}{8} =$ | $\dfrac{8}{6} =$ |
| $\dfrac{12}{6} =$ | $\dfrac{12}{10} =$ | $\dfrac{6}{4} =$ | $\dfrac{20}{12} =$ |
| $\dfrac{5}{3} =$ | $\dfrac{10}{8} =$ | $\dfrac{25}{10} =$ | $\dfrac{10}{3} =$ |
| $\dfrac{15}{10} =$ | $\dfrac{3}{2} =$ | $\dfrac{9}{6} =$ | $\dfrac{7}{4} =$ |
| $\dfrac{18}{12} =$ | $\dfrac{8}{3} =$ | $\dfrac{15}{6} =$ | $\dfrac{14}{4} =$ |
| $\dfrac{8}{4} =$ | $\dfrac{10}{6} =$ | $\dfrac{5}{2} =$ | $\dfrac{21}{12} =$ |
| $\dfrac{15}{12} =$ | $\dfrac{10}{4} =$ | $\dfrac{15}{8} =$ | $\dfrac{4}{3} =$ |

# I

## 28 Improper Fractions to Simplify
*For use with Test 11*

Name _____

Time _____

Write each improper fraction as a mixed number or a whole number.

| | | | |
|---|---|---|---|
| $\frac{5}{4} =$ | $\frac{16}{12} =$ | $\frac{12}{8} =$ | $\frac{8}{6} =$ |
| $\frac{12}{6} =$ | $\frac{12}{10} =$ | $\frac{6}{4} =$ | $\frac{20}{12} =$ |
| $\frac{5}{3} =$ | $\frac{10}{8} =$ | $\frac{25}{10} =$ | $\frac{10}{3} =$ |
| $\frac{15}{10} =$ | $\frac{3}{2} =$ | $\frac{9}{6} =$ | $\frac{7}{4} =$ |
| $\frac{18}{12} =$ | $\frac{8}{3} =$ | $\frac{15}{6} =$ | $\frac{14}{4} =$ |
| $\frac{8}{4} =$ | $\frac{10}{6} =$ | $\frac{5}{2} =$ | $\frac{21}{12} =$ |
| $\frac{15}{12} =$ | $\frac{10}{4} =$ | $\frac{15}{8} =$ | $\frac{4}{3} =$ |

*Saxon Math 7/6—Homeschool*

## H · 72 Multiplication and Division Facts

*For use with Lesson 61*

Name _____

Time _____

Multiply or divide as indicated.

| | | | | | | | |
|---|---|---|---|---|---|---|---|
| 5<br>× 9 | 6)36 | 4<br>× 7 | 8)40 | 10<br>× 3 | 5)20 | 2<br>× 7 | 9)27 |
| 9)81 | 5<br>× 7 | 6)24 | 2<br>× 3 | 7)42 | 7<br>× 9 | 5)10 | 4<br>× 5 |
| 2<br>× 9 | 7)28 | 6<br>× 9 | 4)8 | 3<br>× 3 | 8)24 | 10<br>× 4 | 6)12 |
| 9)63 | 2<br>× 5 | 8)64 | 7<br>× 6 | 10)100 | 3<br>× 5 | 9)36 | 10<br>× 8 |
| 4<br>× 8 | 5)25 | 3<br>× 8 | 4)16 | 10<br>× 5 | 7)14 | 2<br>× 2 | 8)48 |
| 9)54 | 3<br>× 6 | 5)15 | 2<br>× 8 | 7)49 | 4<br>× 6 | 6)30 | 5<br>× 8 |
| 10<br>× 6 | 4)12 | 7<br>× 8 | 6)18 | 3<br>× 4 | 7)35 | 5<br>× 6 | 8)32 |
| 8)56 | 2<br>× 4 | 8)16 | 6<br>× 8 | 7)21 | 8<br>× 9 | 3)6 | 3<br>× 9 |
| 10<br>× 7 | 9)18 | 3<br>× 7 | 9)45 | 2<br>× 6 | 9)72 | 4<br>× 9 | 9)90 |

**G** **30 Fractions to Reduce**
*For use with Lesson 62*

Name _____

Time _____

Reduce each fraction to lowest terms.

| | | | | |
|---|---|---|---|---|
| $\frac{2}{8}$ = | $\frac{4}{6}$ = | $\frac{6}{10}$ = | $\frac{2}{4}$ = | $\frac{6}{16}$ = |
| $\frac{5}{100}$ = | $\frac{9}{12}$ = | $\frac{14}{16}$ = | $\frac{4}{10}$ = | $\frac{4}{12}$ = |
| $\frac{2}{10}$ = | $\frac{3}{6}$ = | $\frac{25}{100}$ = | $\frac{3}{12}$ = | $\frac{4}{16}$ = |
| $\frac{3}{9}$ = | $\frac{10}{16}$ = | $\frac{6}{9}$ = | $\frac{4}{8}$ = | $\frac{2}{12}$ = |
| $\frac{6}{12}$ = | $\frac{2}{16}$ = | $\frac{8}{10}$ = | $\frac{2}{6}$ = | $\frac{75}{100}$ = |
| $\frac{12}{16}$ = | $\frac{8}{12}$ = | $\frac{6}{8}$ = | $\frac{10}{12}$ = | $\frac{5}{10}$ = |

*Saxon Math 7/6—Homeschool*

**64 Multiplication Facts**
*For use with Lesson 63*

Name _____

Time _____

Multiply.

| | | | | | | | |
|---|---|---|---|---|---|---|---|
| 5<br>× 6 | 4<br>× 3 | 9<br>× 8 | 7<br>× 5 | 2<br>× 9 | 8<br>× 4 | 9<br>× 3 | 6<br>× 9 |
| 9<br>× 4 | 2<br>× 5 | 7<br>× 6 | 4<br>× 8 | 7<br>× 9 | 5<br>× 4 | 3<br>× 2 | 9<br>× 7 |
| 3<br>× 7 | 8<br>× 5 | 6<br>× 2 | 5<br>× 5 | 3<br>× 5 | 2<br>× 4 | 7<br>× 7 | 8<br>× 9 |
| 6<br>× 4 | 2<br>× 8 | 4<br>× 4 | 8<br>× 2 | 3<br>× 9 | 6<br>× 6 | 9<br>× 9 | 5<br>× 3 |
| 4<br>× 6 | 8<br>× 8 | 5<br>× 7 | 6<br>× 3 | 2<br>× 2 | 7<br>× 4 | 3<br>× 8 | 8<br>× 6 |
| 2<br>× 6 | 5<br>× 9 | 3<br>× 3 | 9<br>× 2 | 6<br>× 7 | 4<br>× 5 | 7<br>× 2 | 9<br>× 6 |
| 5<br>× 2 | 7<br>× 8 | 2<br>× 3 | 6<br>× 8 | 4<br>× 7 | 9<br>× 5 | 3<br>× 6 | 8<br>× 7 |
| 3<br>× 4 | 7<br>× 3 | 5<br>× 8 | 4<br>× 2 | 8<br>× 3 | 2<br>× 7 | 6<br>× 5 | 4<br>× 9 |

J

**24 Mixed Numbers to Write as Improper Fractions**

*For use with Lesson 64*

Name _____

Time _____

Write each mixed number as an improper fraction.

| | | | |
|---|---|---|---|
| $2\frac{1}{2} =$ | $2\frac{2}{5} =$ | $1\frac{3}{4} =$ | $2\frac{3}{4} =$ |
| $2\frac{1}{8} =$ | $1\frac{2}{3} =$ | $10\frac{1}{2} =$ | $2\frac{1}{3} =$ |
| $3\frac{1}{2} =$ | $1\frac{5}{6} =$ | $2\frac{1}{4} =$ | $1\frac{1}{8} =$ |
| $5\frac{1}{2} =$ | $1\frac{3}{8} =$ | $5\frac{1}{3} =$ | $3\frac{1}{4} =$ |
| $4\frac{1}{2} =$ | $1\frac{7}{8} =$ | $2\frac{2}{3} =$ | $3\frac{3}{10} =$ |
| $1\frac{5}{8} =$ | $3\frac{3}{4} =$ | $2\frac{3}{8} =$ | $7\frac{1}{2} =$ |

*Saxon Math 7/6—Homeschool*

## H · 72 Multiplication and Division Facts

*For use with Lesson 65*

Name _____

Time _____

Multiply or divide as indicated.

| | | | | | | |
|---|---|---|---|---|---|---|
| $\times\ \begin{array}{r}5\\9\end{array}$ | $6\overline{)36}$ | $\times\ \begin{array}{r}4\\7\end{array}$ | $8\overline{)40}$ | $\times\ \begin{array}{r}10\\3\end{array}$ | $5\overline{)20}$ | $\times\ \begin{array}{r}2\\7\end{array}$ | $9\overline{)27}$ |
| $9\overline{)81}$ | $\times\ \begin{array}{r}5\\7\end{array}$ | $6\overline{)24}$ | $\times\ \begin{array}{r}2\\3\end{array}$ | $7\overline{)42}$ | $\times\ \begin{array}{r}7\\9\end{array}$ | $5\overline{)10}$ | $\times\ \begin{array}{r}4\\5\end{array}$ |
| $\times\ \begin{array}{r}2\\9\end{array}$ | $7\overline{)28}$ | $\times\ \begin{array}{r}6\\9\end{array}$ | $4\overline{)8}$ | $\times\ \begin{array}{r}3\\3\end{array}$ | $8\overline{)24}$ | $\times\ \begin{array}{r}10\\4\end{array}$ | $6\overline{)12}$ |
| $9\overline{)63}$ | $\times\ \begin{array}{r}2\\5\end{array}$ | $8\overline{)64}$ | $\times\ \begin{array}{r}7\\6\end{array}$ | $10\overline{)100}$ | $\times\ \begin{array}{r}3\\5\end{array}$ | $9\overline{)36}$ | $\times\ \begin{array}{r}10\\8\end{array}$ |
| $\times\ \begin{array}{r}4\\8\end{array}$ | $5\overline{)25}$ | $\times\ \begin{array}{r}3\\8\end{array}$ | $4\overline{)16}$ | $\times\ \begin{array}{r}10\\5\end{array}$ | $7\overline{)14}$ | $\times\ \begin{array}{r}2\\2\end{array}$ | $8\overline{)48}$ |
| $9\overline{)54}$ | $\times\ \begin{array}{r}3\\6\end{array}$ | $5\overline{)15}$ | $\times\ \begin{array}{r}2\\8\end{array}$ | $7\overline{)49}$ | $\times\ \begin{array}{r}4\\6\end{array}$ | $6\overline{)30}$ | $\times\ \begin{array}{r}5\\8\end{array}$ |
| $\times\ \begin{array}{r}10\\6\end{array}$ | $4\overline{)12}$ | $\times\ \begin{array}{r}7\\8\end{array}$ | $6\overline{)18}$ | $\times\ \begin{array}{r}3\\4\end{array}$ | $7\overline{)35}$ | $\times\ \begin{array}{r}5\\6\end{array}$ | $8\overline{)32}$ |
| $8\overline{)56}$ | $\times\ \begin{array}{r}2\\4\end{array}$ | $8\overline{)16}$ | $\times\ \begin{array}{r}6\\8\end{array}$ | $7\overline{)21}$ | $\times\ \begin{array}{r}8\\9\end{array}$ | $3\overline{)6}$ | $\times\ \begin{array}{r}3\\9\end{array}$ |
| $\times\ \begin{array}{r}10\\7\end{array}$ | $9\overline{)18}$ | $\times\ \begin{array}{r}3\\7\end{array}$ | $9\overline{)45}$ | $\times\ \begin{array}{r}2\\6\end{array}$ | $9\overline{)72}$ | $\times\ \begin{array}{r}4\\9\end{array}$ | $9\overline{)90}$ |

<table>
<tr><td>H</td><td>72 Multiplication and<br>Division Facts<br>For use with Test 12</td><td>Name _____<br><br>Time _____</td></tr>
</table>

Multiply or divide as indicated.

| | | | | | | | |
|---|---|---|---|---|---|---|---|
| 5<br>× 9 | 6)36 | 4<br>× 7 | 8)40 | 10<br>× 3 | 5)20 | 2<br>× 7 | 9)27 |
| 9)81 | 5<br>× 7 | 6)24 | 2<br>× 3 | 7)42 | 7<br>× 9 | 5)10 | 4<br>× 5 |
| 2<br>× 9 | 7)28 | 6<br>× 9 | 4)8 | 3<br>× 3 | 8)24 | 10<br>× 4 | 6)12 |
| 9)63 | 2<br>× 5 | 8)64 | 7<br>× 6 | 10)100 | 3<br>× 5 | 9)36 | 10<br>× 8 |
| 4<br>× 8 | 5)25 | 3<br>× 8 | 4)16 | 10<br>× 5 | 7)14 | 2<br>× 2 | 8)48 |
| 9)54 | 3<br>× 6 | 5)15 | 2<br>× 8 | 7)49 | 4<br>× 6 | 6)30 | 5<br>× 8 |
| 10<br>× 6 | 4)12 | 7<br>× 8 | 6)18 | 3<br>× 4 | 7)35 | 5<br>× 6 | 8)32 |
| 8)56 | 2<br>× 4 | 8)16 | 6<br>× 8 | 7)21 | 8<br>× 9 | 3)6 | 3<br>× 9 |
| 10<br>× 7 | 9)18 | 3<br>× 7 | 9)45 | 2<br>× 6 | 9)72 | 4<br>× 9 | 9)90 |

Saxon Math 7/6—Homeschool

# J

## 24 Mixed Numbers to Write as Improper Fractions

*For use with Lesson 66*

Name _____

Time _____

Write each mixed number as an improper fraction.

| | | | |
|---|---|---|---|
| $2\frac{1}{2} =$ | $2\frac{2}{5} =$ | $1\frac{3}{4} =$ | $2\frac{3}{4} =$ |
| $2\frac{1}{8} =$ | $1\frac{2}{3} =$ | $10\frac{1}{2} =$ | $2\frac{1}{3} =$ |
| $3\frac{1}{2} =$ | $1\frac{5}{6} =$ | $2\frac{1}{4} =$ | $1\frac{1}{8} =$ |
| $5\frac{1}{2} =$ | $1\frac{3}{8} =$ | $5\frac{1}{3} =$ | $3\frac{1}{4} =$ |
| $4\frac{1}{2} =$ | $1\frac{7}{8} =$ | $2\frac{2}{3} =$ | $3\frac{3}{10} =$ |
| $1\frac{5}{8} =$ | $3\frac{3}{4} =$ | $2\frac{3}{8} =$ | $7\frac{1}{2} =$ |

## J

### 24 Mixed Numbers to Write as Improper Fractions

*For use with Lesson 67*

Name _____

Time _____

Write each mixed number as an improper fraction.

| | | | |
|---|---|---|---|
| $2\frac{1}{2} =$ | $2\frac{2}{5} =$ | $1\frac{3}{4} =$ | $2\frac{3}{4} =$ |
| $2\frac{1}{8} =$ | $1\frac{2}{3} =$ | $10\frac{1}{2} =$ | $2\frac{1}{3} =$ |
| $3\frac{1}{2} =$ | $1\frac{5}{6} =$ | $2\frac{1}{4} =$ | $1\frac{1}{8} =$ |
| $5\frac{1}{2} =$ | $1\frac{3}{8} =$ | $5\frac{1}{3} =$ | $3\frac{1}{4} =$ |
| $4\frac{1}{2} =$ | $1\frac{7}{8} =$ | $2\frac{2}{3} =$ | $3\frac{3}{10} =$ |
| $1\frac{5}{8} =$ | $3\frac{3}{4} =$ | $2\frac{3}{8} =$ | $7\frac{1}{2} =$ |

## I

## 28 Improper Fractions to Simplify

*For use with Lesson 68*

Name _____

Time _____

Write each improper fraction as a mixed number or a whole number.

| | | | |
|---|---|---|---|
| $\frac{5}{4} =$ | $\frac{16}{12} =$ | $\frac{12}{8} =$ | $\frac{8}{6} =$ |
| $\frac{12}{6} =$ | $\frac{12}{10} =$ | $\frac{6}{4} =$ | $\frac{20}{12} =$ |
| $\frac{5}{3} =$ | $\frac{10}{8} =$ | $\frac{25}{10} =$ | $\frac{10}{3} =$ |
| $\frac{15}{10} =$ | $\frac{3}{2} =$ | $\frac{9}{6} =$ | $\frac{7}{4} =$ |
| $\frac{18}{12} =$ | $\frac{8}{3} =$ | $\frac{15}{6} =$ | $\frac{14}{4} =$ |
| $\frac{8}{4} =$ | $\frac{10}{6} =$ | $\frac{5}{2} =$ | $\frac{21}{12} =$ |
| $\frac{15}{12} =$ | $\frac{10}{4} =$ | $\frac{15}{8} =$ | $\frac{4}{3} =$ |

# J

## 24 Mixed Numbers to Write as Improper Fractions

*For use with Lesson 69*

Name _____

Time _____

Write each mixed number as an improper fraction.

| | | | |
|---|---|---|---|
| $2\frac{1}{2} =$ | $2\frac{2}{5} =$ | $1\frac{3}{4} =$ | $2\frac{3}{4} =$ |
| $2\frac{1}{8} =$ | $1\frac{2}{3} =$ | $10\frac{1}{2} =$ | $2\frac{1}{3} =$ |
| $3\frac{1}{2} =$ | $1\frac{5}{6} =$ | $2\frac{1}{4} =$ | $1\frac{1}{8} =$ |
| $5\frac{1}{2} =$ | $1\frac{3}{8} =$ | $5\frac{1}{3} =$ | $3\frac{1}{4} =$ |
| $4\frac{1}{2} =$ | $1\frac{7}{8} =$ | $2\frac{2}{3} =$ | $3\frac{3}{10} =$ |
| $1\frac{5}{8} =$ | $3\frac{3}{4} =$ | $2\frac{3}{8} =$ | $7\frac{1}{2} =$ |

*Saxon Math 7/6—Homeschool*

## G | 30 Fractions to Reduce

*For use with Lesson 70*

Name _____

Time _____

Reduce each fraction to lowest terms.

| | | | | |
|---|---|---|---|---|
| $\dfrac{2}{8} =$ | $\dfrac{4}{6} =$ | $\dfrac{6}{10} =$ | $\dfrac{2}{4} =$ | $\dfrac{6}{16} =$ |
| $\dfrac{5}{100} =$ | $\dfrac{9}{12} =$ | $\dfrac{14}{16} =$ | $\dfrac{4}{10} =$ | $\dfrac{4}{12} =$ |
| $\dfrac{2}{10} =$ | $\dfrac{3}{6} =$ | $\dfrac{25}{100} =$ | $\dfrac{3}{12} =$ | $\dfrac{4}{16} =$ |
| $\dfrac{3}{9} =$ | $\dfrac{10}{16} =$ | $\dfrac{6}{9} =$ | $\dfrac{4}{8} =$ | $\dfrac{2}{12} =$ |
| $\dfrac{6}{12} =$ | $\dfrac{2}{16} =$ | $\dfrac{8}{10} =$ | $\dfrac{2}{6} =$ | $\dfrac{75}{100} =$ |
| $\dfrac{12}{16} =$ | $\dfrac{8}{12} =$ | $\dfrac{6}{8} =$ | $\dfrac{10}{12} =$ | $\dfrac{5}{10} =$ |

J

**24 Mixed Numbers to Write as Improper Fractions**

*For use with Test 13*

Name _____

Time _____

Write each mixed number as an improper fraction.

| | | | |
|---|---|---|---|
| $2\frac{1}{2} =$ | $2\frac{2}{5} =$ | $1\frac{3}{4} =$ | $2\frac{3}{4} =$ |
| $2\frac{1}{8} =$ | $1\frac{2}{3} =$ | $10\frac{1}{2} =$ | $2\frac{1}{3} =$ |
| $3\frac{1}{2} =$ | $1\frac{5}{6} =$ | $2\frac{1}{4} =$ | $1\frac{1}{8} =$ |
| $5\frac{1}{2} =$ | $1\frac{3}{8} =$ | $5\frac{1}{3} =$ | $3\frac{1}{4} =$ |
| $4\frac{1}{2} =$ | $1\frac{7}{8} =$ | $2\frac{2}{3} =$ | $3\frac{3}{10} =$ |
| $1\frac{5}{8} =$ | $3\frac{3}{4} =$ | $2\frac{3}{8} =$ | $7\frac{1}{2} =$ |

D

**64 Multiplication Facts**
*For use with Lesson 71*

Name _____

Time _____

Multiply.

| | | | | | | | |
|---|---|---|---|---|---|---|---|
| 5<br>× 6 | 4<br>× 3 | 9<br>× 8 | 7<br>× 5 | 2<br>× 9 | 8<br>× 4 | 9<br>× 3 | 6<br>× 9 |
| 9<br>× 4 | 2<br>× 5 | 7<br>× 6 | 4<br>× 8 | 7<br>× 9 | 5<br>× 4 | 3<br>× 2 | 9<br>× 7 |
| 3<br>× 7 | 8<br>× 5 | 6<br>× 2 | 5<br>× 5 | 3<br>× 5 | 2<br>× 4 | 7<br>× 7 | 8<br>× 9 |
| 6<br>× 4 | 2<br>× 8 | 4<br>× 4 | 8<br>× 2 | 3<br>× 9 | 6<br>× 6 | 9<br>× 9 | 5<br>× 3 |
| 4<br>× 6 | 8<br>× 8 | 5<br>× 7 | 6<br>× 3 | 2<br>× 2 | 7<br>× 4 | 3<br>× 8 | 8<br>× 6 |
| 2<br>× 6 | 5<br>× 9 | 3<br>× 3 | 9<br>× 2 | 6<br>× 7 | 4<br>× 5 | 7<br>× 2 | 9<br>× 6 |
| 5<br>× 2 | 7<br>× 8 | 2<br>× 3 | 6<br>× 8 | 4<br>× 7 | 9<br>× 5 | 3<br>× 6 | 8<br>× 7 |
| 3<br>× 4 | 7<br>× 3 | 5<br>× 8 | 4<br>× 2 | 8<br>× 3 | 2<br>× 7 | 6<br>× 5 | 4<br>× 9 |

| H |
|---|

## 72 Multiplication and Division Facts
*For use with Lesson 72*

Name _____

Time _____

Multiply or divide as indicated.

| | | | | | | | |
|---|---|---|---|---|---|---|---|
| $\times\ \begin{array}{r}5\\9\end{array}$ | $6\overline{)36}$ | $\times\ \begin{array}{r}4\\7\end{array}$ | $8\overline{)40}$ | $\times\ \begin{array}{r}10\\3\end{array}$ | $5\overline{)20}$ | $\times\ \begin{array}{r}2\\7\end{array}$ | $9\overline{)27}$ |
| $9\overline{)81}$ | $\times\ \begin{array}{r}5\\7\end{array}$ | $6\overline{)24}$ | $\times\ \begin{array}{r}2\\3\end{array}$ | $7\overline{)42}$ | $\times\ \begin{array}{r}7\\9\end{array}$ | $5\overline{)10}$ | $\times\ \begin{array}{r}4\\5\end{array}$ |
| $\times\ \begin{array}{r}2\\9\end{array}$ | $7\overline{)28}$ | $\times\ \begin{array}{r}6\\9\end{array}$ | $4\overline{)8}$ | $\times\ \begin{array}{r}3\\3\end{array}$ | $8\overline{)24}$ | $\times\ \begin{array}{r}10\\4\end{array}$ | $6\overline{)12}$ |
| $9\overline{)63}$ | $\times\ \begin{array}{r}2\\5\end{array}$ | $8\overline{)64}$ | $\times\ \begin{array}{r}7\\6\end{array}$ | $10\overline{)100}$ | $\times\ \begin{array}{r}3\\5\end{array}$ | $9\overline{)36}$ | $\times\ \begin{array}{r}10\\8\end{array}$ |
| $\times\ \begin{array}{r}4\\8\end{array}$ | $5\overline{)25}$ | $\times\ \begin{array}{r}3\\8\end{array}$ | $4\overline{)16}$ | $\times\ \begin{array}{r}10\\5\end{array}$ | $7\overline{)14}$ | $\times\ \begin{array}{r}2\\2\end{array}$ | $8\overline{)48}$ |
| $9\overline{)54}$ | $\times\ \begin{array}{r}3\\6\end{array}$ | $5\overline{)15}$ | $\times\ \begin{array}{r}2\\8\end{array}$ | $7\overline{)49}$ | $\times\ \begin{array}{r}4\\6\end{array}$ | $6\overline{)30}$ | $\times\ \begin{array}{r}5\\8\end{array}$ |
| $\times\ \begin{array}{r}10\\6\end{array}$ | $4\overline{)12}$ | $\times\ \begin{array}{r}7\\8\end{array}$ | $6\overline{)18}$ | $\times\ \begin{array}{r}3\\4\end{array}$ | $7\overline{)35}$ | $\times\ \begin{array}{r}5\\6\end{array}$ | $8\overline{)32}$ |
| $8\overline{)56}$ | $\times\ \begin{array}{r}2\\4\end{array}$ | $8\overline{)16}$ | $\times\ \begin{array}{r}6\\8\end{array}$ | $7\overline{)21}$ | $\times\ \begin{array}{r}8\\9\end{array}$ | $3\overline{)6}$ | $\times\ \begin{array}{r}3\\9\end{array}$ |
| $\times\ \begin{array}{r}10\\7\end{array}$ | $9\overline{)18}$ | $\times\ \begin{array}{r}3\\7\end{array}$ | $9\overline{)45}$ | $\times\ \begin{array}{r}2\\6\end{array}$ | $9\overline{)72}$ | $\times\ \begin{array}{r}4\\9\end{array}$ | $9\overline{)90}$ |

*Saxon Math 7/6—Homeschool*

# J

## 24 Mixed Numbers to Write as Improper Fractions

*For use with Lesson 73*

Name _____

Time _____

Write each mixed number as an improper fraction.

| | | | |
|---|---|---|---|
| $2\dfrac{1}{2} =$ | $2\dfrac{2}{5} =$ | $1\dfrac{3}{4} =$ | $2\dfrac{3}{4} =$ |
| $2\dfrac{1}{8} =$ | $1\dfrac{2}{3} =$ | $10\dfrac{1}{2} =$ | $2\dfrac{1}{3} =$ |
| $3\dfrac{1}{2} =$ | $1\dfrac{5}{6} =$ | $2\dfrac{1}{4} =$ | $1\dfrac{1}{8} =$ |
| $5\dfrac{1}{2} =$ | $1\dfrac{3}{8} =$ | $5\dfrac{1}{3} =$ | $3\dfrac{1}{4} =$ |
| $4\dfrac{1}{2} =$ | $1\dfrac{7}{8} =$ | $2\dfrac{2}{3} =$ | $3\dfrac{3}{10} =$ |
| $1\dfrac{5}{8} =$ | $3\dfrac{3}{4} =$ | $2\dfrac{3}{8} =$ | $7\dfrac{1}{2} =$ |

## I

## 28 Improper Fractions to Simplify

*For use with Lesson 74*

Name _____

Time _____

Write each improper fraction as a mixed number or a whole number.

| | | | |
|---|---|---|---|
| $\dfrac{5}{4} =$ | $\dfrac{16}{12} =$ | $\dfrac{12}{8} =$ | $\dfrac{8}{6} =$ |
| $\dfrac{12}{6} =$ | $\dfrac{12}{10} =$ | $\dfrac{6}{4} =$ | $\dfrac{20}{12} =$ |
| $\dfrac{5}{3} =$ | $\dfrac{10}{8} =$ | $\dfrac{25}{10} =$ | $\dfrac{10}{3} =$ |
| $\dfrac{15}{10} =$ | $\dfrac{3}{2} =$ | $\dfrac{9}{6} =$ | $\dfrac{7}{4} =$ |
| $\dfrac{18}{12} =$ | $\dfrac{8}{3} =$ | $\dfrac{15}{6} =$ | $\dfrac{14}{4} =$ |
| $\dfrac{8}{4} =$ | $\dfrac{10}{6} =$ | $\dfrac{5}{2} =$ | $\dfrac{21}{12} =$ |
| $\dfrac{15}{12} =$ | $\dfrac{10}{4} =$ | $\dfrac{15}{8} =$ | $\dfrac{4}{3} =$ |

*Saxon Math 7/6—Homeschool*

## K

## Linear Measurement
*For use with Lesson 75*

Name _____

Time _____

---

Write the abbreviation for each of the following units.

Metric Units

1. millimeter _____
2. centimeter _____
3. meter _____
4. kilometer _____

U.S. Customary Units

5. inch _____
6. foot _____
7. yard _____
8. mile _____

---

Complete each unit conversion.

Metric Conversions

9. 1 centimeter = _____ millimeters
10. 1 meter = _____ millimeters
11. 1 meter = _____ centimeters
12. 1 kilometer = _____ meters

U.S. Customary Conversions

13. 1 foot = _____ inches
14. 1 yard = _____ inches
15. 1 yard = _____ feet
16. 1 mile = _____ feet
17. 1 mile = _____ yards

Conversions between systems

18. 1 inch = _____ centimeters
19. 1 mile ≈ _____ meters

20. 1 meter ≈ _____ inches
21. 1 kilometer ≈ _____ mile

---

Write an appropriate unit for each physical reference.

Metric Units

22. The thickness of a dime: _____
23. The width of a little finger: _____
24. The length of one BIG step: _____

U.S. Customary Units

25. The width of two fingers: _____
26. The length of a man's shoe: _____
27. The length of one big step: _____

---

Arrange each set of units in order from shortest to longest.

28. m, cm, mm, km _____, _____, _____, _____
29. ft, mi, in., yd _____, _____, _____, _____

---

Find each equivalent measure.

30. 10 cm = _____ mm
31. 2 m = _____ cm or _____ mm
32. 5 km = _____ m
33. 2.5 cm = _____ mm
34. 1.5 m = _____ cm or _____ mm
35. 7.5 km = _____ m

36. $\frac{1}{2}$ ft = _____ in.
37. 2 ft = _____ in.
38. 3 ft = _____ in.
39. 2 yd = _____ ft
40. 10 yd = _____ ft
41. 100 yd = _____ ft

---

## 28 Improper Fractions to Simplify

*For use with Test 14*

Name _____

Time _____

Write each improper fraction as a mixed number or a whole number.

| | | | |
|---|---|---|---|
| $\dfrac{5}{4} =$ | $\dfrac{16}{12} =$ | $\dfrac{12}{8} =$ | $\dfrac{8}{6} =$ |
| $\dfrac{12}{6} =$ | $\dfrac{12}{10} =$ | $\dfrac{6}{4} =$ | $\dfrac{20}{12} =$ |
| $\dfrac{5}{3} =$ | $\dfrac{10}{8} =$ | $\dfrac{25}{10} =$ | $\dfrac{10}{3} =$ |
| $\dfrac{15}{10} =$ | $\dfrac{3}{2} =$ | $\dfrac{9}{6} =$ | $\dfrac{7}{4} =$ |
| $\dfrac{18}{12} =$ | $\dfrac{8}{3} =$ | $\dfrac{15}{6} =$ | $\dfrac{14}{4} =$ |
| $\dfrac{8}{4} =$ | $\dfrac{10}{6} =$ | $\dfrac{5}{2} =$ | $\dfrac{21}{12} =$ |
| $\dfrac{15}{12} =$ | $\dfrac{10}{4} =$ | $\dfrac{15}{8} =$ | $\dfrac{4}{3} =$ |

*Saxon Math 7/6—Homeschool*

## 30 Fractions to Reduce

*For use with Lesson 76*

Name _____

Time _____

Reduce each fraction to lowest terms.

| | | | | |
|---|---|---|---|---|
| $\frac{2}{8} =$ | $\frac{4}{6} =$ | $\frac{6}{10} =$ | $\frac{2}{4} =$ | $\frac{6}{16} =$ |
| $\frac{5}{100} =$ | $\frac{9}{12} =$ | $\frac{14}{16} =$ | $\frac{4}{10} =$ | $\frac{4}{12} =$ |
| $\frac{2}{10} =$ | $\frac{3}{6} =$ | $\frac{25}{100} =$ | $\frac{3}{12} =$ | $\frac{4}{16} =$ |
| $\frac{3}{9} =$ | $\frac{10}{16} =$ | $\frac{6}{9} =$ | $\frac{4}{8} =$ | $\frac{2}{12} =$ |
| $\frac{6}{12} =$ | $\frac{2}{16} =$ | $\frac{8}{10} =$ | $\frac{2}{6} =$ | $\frac{75}{100} =$ |
| $\frac{12}{16} =$ | $\frac{8}{12} =$ | $\frac{6}{8} =$ | $\frac{10}{12} =$ | $\frac{5}{10} =$ |

## K — Linear Measurement
*For use with Lesson 77*

Name _____

Time _____

Write the abbreviation for each of the following units.

Metric Units

1. millimeter _____
2. centimeter _____
3. meter _____
4. kilometer _____

U.S. Customary Units

5. inch _____
6. foot _____
7. yard _____
8. mile _____

Complete each unit conversion.

Metric Conversions

9. 1 centimeter = _____ millimeters
10. 1 meter = _____ millimeters
11. 1 meter = _____ centimeters
12. 1 kilometer = _____ meters

U.S. Customary Conversions

13. 1 foot = _____ inches
14. 1 yard = _____ inches
15. 1 yard = _____ feet
16. 1 mile = _____ feet
17. 1 mile = _____ yards

Conversions between systems

18. 1 inch = _____ centimeters
19. 1 mile ≈ _____ meters
20. 1 meter ≈ _____ inches
21. 1 kilometer ≈ _____ mile

Write an appropriate unit for each physical reference.

Metric Units

22. The thickness of a dime: _____
23. The width of a little finger: _____
24. The length of one BIG step: _____

U.S. Customary Units

25. The width of two fingers: _____
26. The length of a man's shoe: _____
27. The length of one big step: _____

Arrange each set of units in order from shortest to longest.

28. m, cm, mm, km _____, _____, _____, _____
29. ft, mi, in., yd _____, _____, _____, _____

Find each equivalent measure.

30. 10 cm = _____ mm
31. 2 m = _____ cm or _____ mm
32. 5 km = _____ m
33. 2.5 cm = _____ mm
34. 1.5 m = _____ cm or _____ mm
35. 7.5 km = _____ m
36. $\frac{1}{2}$ ft = _____ in.
37. 2 ft = _____ in.
38. 3 ft = _____ in.
39. 2 yd = _____ ft
40. 10 yd = _____ ft
41. 100 yd = _____ ft

**J**

## 24 Mixed Numbers to Write as Improper Fractions

*For use with Lesson 78*

Name _____

Time _____

Write each mixed number as an improper fraction.

| | | | |
|---|---|---|---|
| $2\frac{1}{2} =$ | $2\frac{2}{5} =$ | $1\frac{3}{4} =$ | $2\frac{3}{4} =$ |
| $2\frac{1}{8} =$ | $1\frac{2}{3} =$ | $10\frac{1}{2} =$ | $2\frac{1}{3} =$ |
| $3\frac{1}{2} =$ | $1\frac{5}{6} =$ | $2\frac{1}{4} =$ | $1\frac{1}{8} =$ |
| $5\frac{1}{2} =$ | $1\frac{3}{8} =$ | $5\frac{1}{3} =$ | $3\frac{1}{4} =$ |
| $4\frac{1}{2} =$ | $1\frac{7}{8} =$ | $2\frac{2}{3} =$ | $3\frac{3}{10} =$ |
| $1\frac{5}{8} =$ | $3\frac{3}{4} =$ | $2\frac{3}{8} =$ | $7\frac{1}{2} =$ |

## K Linear Measurement
*For use with Lesson 79*

Name _____

Time _____

Write the abbreviation for each of the following units.

Metric Units

1. millimeter _____
2. centimeter _____
3. meter _____
4. kilometer _____

U.S. Customary Units

5. inch _____
6. foot _____
7. yard _____
8. mile _____

Complete each unit conversion.

Metric Conversions

9. 1 centimeter = _____ millimeters
10. 1 meter = _____ millimeters
11. 1 meter = _____ centimeters
12. 1 kilometer = _____ meters

U.S. Customary Conversions

13. 1 foot = _____ inches
14. 1 yard = _____ inches
15. 1 yard = _____ feet
16. 1 mile = _____ feet
17. 1 mile = _____ yards

Conversions between systems

18. 1 inch = _____ centimeters
19. 1 mile ≈ _____ meters
20. 1 meter ≈ _____ inches
21. 1 kilometer ≈ _____ mile

Write an appropriate unit for each physical reference.

Metric Units

22. The thickness of a dime: _____
23. The width of a little finger: _____
24. The length of one BIG step: _____

U.S. Customary Units

25. The width of two fingers: _____
26. The length of a man's shoe: _____
27. The length of one big step: _____

Arrange each set of units in order from shortest to longest.

28. m, cm, mm, km _____, _____, _____, _____
29. ft, mi, in., yd _____, _____, _____, _____

Find each equivalent measure.

30. 10 cm = _____ mm
31. 2 m = _____ cm or _____ mm
32. 5 km = _____ m
33. 2.5 cm = _____ mm
34. 1.5 m = _____ cm or _____ mm
35. 7.5 km = _____ m

36. $\frac{1}{2}$ ft = _____ in.
37. 2 ft = _____ in.
38. 3 ft = _____ in.
39. 2 yd = _____ ft
40. 10 yd = _____ ft
41. 100 yd = _____ ft

*Saxon Math 7/6—Homeschool*

## 16 | Constructing Bisectors
*For use with Investigation 8*

Name _____

**Section A**

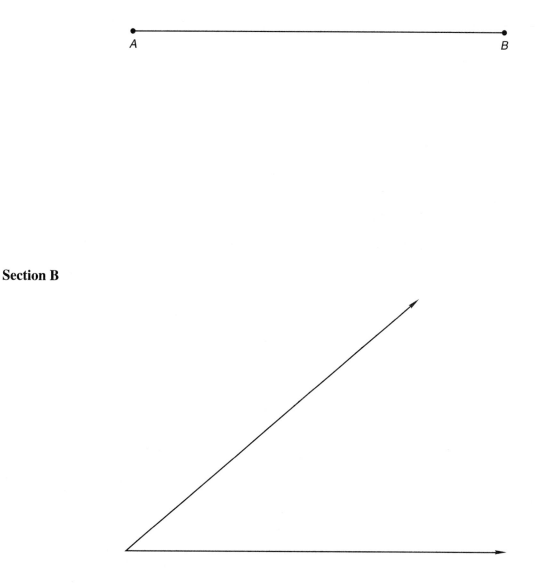

A                                              B

**Section B**

## K — Linear Measurement
*For use with Lesson 81*

Name _____

Time _____

---

Write the abbreviation for each of the following units.

Metric Units

**1.** millimeter _____

**2.** centimeter _____

**3.** meter _____

**4.** kilometer _____

U.S. Customary Units

**5.** inch _____

**6.** foot _____

**7.** yard _____

**8.** mile _____

---

Complete each unit conversion.

Metric Conversions

**9.** 1 centimeter = _____ millimeters

**10.** 1 meter = _____ millimeters

**11.** 1 meter = _____ centimeters

**12.** 1 kilometer = _____ meters

U.S. Customary Conversions

**13.** 1 foot = _____ inches

**14.** 1 yard = _____ inches

**15.** 1 yard = _____ feet

**16.** 1 mile = _____ feet

**17.** 1 mile = _____ yards

Conversions between systems

**18.** 1 inch = _____ centimeters

**19.** 1 mile ≈ _____ meters

**20.** 1 meter ≈ _____ inches

**21.** 1 kilometer ≈ _____ mile

---

Write an appropriate unit for each physical reference.

Metric Units

**22.** The thickness of a dime: _____

**23.** The width of a little finger: _____

**24.** The length of one BIG step: _____

U.S. Customary Units

**25.** The width of two fingers: _____

**26.** The length of a man's shoe: _____

**27.** The length of one big step: _____

---

Arrange each set of units in order from shortest to longest.

**28.** m, cm, mm, km _____, _____, _____, _____

**29.** ft, mi, in., yd _____, _____, _____, _____

---

Find each equivalent measure.

**30.** 10 cm = _____ mm

**31.** 2 m = _____ cm or _____ mm

**32.** 5 km = _____ m

**33.** 2.5 cm = _____ mm

**34.** 1.5 m = _____ cm or _____ mm

**35.** 7.5 km = _____ m

**36.** $\frac{1}{2}$ ft = _____ in.

**37.** 2 ft = _____ in.

**38.** 3 ft = _____ in.

**39.** 2 yd = _____ ft

**40.** 10 yd = _____ ft

**41.** 100 yd = _____ ft

---

## L | Liquid Measurement
*For use with Lesson 82*

Name _____

Time _____

---

Write the abbreviation for each of the following units.

Metric Units                              U.S. Customary Units

**1.** liter   _____        **3.** ounce _____        **5.** quart   _____

**2.** milliliter _____        **4.** pint   _____        **6.** gallon _____

---

Complete each unit conversion.

Metric Conversions                        U.S. Customary Conversions

**7.** 1 liter = _____ milliliters        **8.** 1 cup   = _____ ounces

                                                  **9.** 1 pint  = _____ ounces

                                                  **10.** 1 quart = _____ pints

                                                  **11.** 1 gallon = _____ quarts

Conversions between systems

**12.** 1 liter ≈ _____ quarts        **13.** 1 gallon ≈ _____ liters

---

Complete each statement.

**14.** One fourth of a dollar is a quarter, and one fourth of a gallon is a _____.

**15.** A two-liter bottle of soda is a little more than 2 _____ or $\frac{1}{2}$ of a _____.

**16.** "A pint's a pound the world around" means a _____ of water weighs about 1 pound.

**17.** In the United States, gasoline is sold by the _____. In other countries, gasoline is

sold by the _____. A gallon of gasoline is more than _____ liters but a little

less than _____ liters.

**18.** 2 cups make a _____.

**19.** 2 pints make a _____.

**20.** 2 quarts make a _____.

**21.** 2 half gallons make a _____.

**22.** A gallon of milk will fill _____ cups.

**23.** If you drink 8 cups of water each day, you drink a _____ of water.

---

Find each equivalent measure.

**24.** 2 liters   = _____ milliliters        **28.** $\frac{1}{2}$ pint (1 cup)   = _____ ounces

**25.** 2 liters   ≈ _____ quarts        **29.** 1 quart (2 pints) = _____ ounces

**26.** 3.78 liters = _____ milliliters        **30.** $\frac{1}{2}$ gallon   = _____ quarts

**27.** 0.5 liters = _____ milliliters        **31.** 1 gallon   = _____ pints

                                                  **32.** 2 gallons   = _____ quarts

## I   28 Improper Fractions to Simplify

*For use with Lesson 83*

Name _____

Time _____

Write each improper fraction as a mixed number or a whole number.

| | | | |
|---|---|---|---|
| $\frac{5}{4} =$ | $\frac{16}{12} =$ | $\frac{12}{8} =$ | $\frac{8}{6} =$ |
| $\frac{12}{6} =$ | $\frac{12}{10} =$ | $\frac{6}{4} =$ | $\frac{20}{12} =$ |
| $\frac{5}{3} =$ | $\frac{10}{8} =$ | $\frac{25}{10} =$ | $\frac{10}{3} =$ |
| $\frac{15}{10} =$ | $\frac{3}{2} =$ | $\frac{9}{6} =$ | $\frac{7}{4} =$ |
| $\frac{18}{12} =$ | $\frac{8}{3} =$ | $\frac{15}{6} =$ | $\frac{14}{4} =$ |
| $\frac{8}{4} =$ | $\frac{10}{6} =$ | $\frac{5}{2} =$ | $\frac{21}{12} =$ |
| $\frac{15}{12} =$ | $\frac{10}{4} =$ | $\frac{15}{8} =$ | $\frac{4}{3} =$ |

# K Linear Measurement

*For use with Lesson 84*

Name _____

Time _____

---

Write the abbreviation for each of the following units.

Metric Units

1. millimeter _____

2. centimeter _____

3. meter _____

4. kilometer _____

U.S. Customary Units

5. inch _____

6. foot _____

7. yard _____

8. mile _____

---

Complete each unit conversion.

Metric Conversions

9. 1 centimeter = _____ millimeters

10. 1 meter = _____ millimeters

11. 1 meter = _____ centimeters

12. 1 kilometer = _____ meters

U.S. Customary Conversions

13. 1 foot = _____ inches

14. 1 yard = _____ inches

15. 1 yard = _____ feet

16. 1 mile = _____ feet

17. 1 mile = _____ yards

Conversions between systems

18. 1 inch = _____ centimeters

19. 1 mile ≈ _____ meters

20. 1 meter ≈ _____ inches

21. 1 kilometer ≈ _____ mile

---

Write an appropriate unit for each physical reference.

Metric Units

22. The thickness of a dime: _____

23. The width of a little finger: _____

24. The length of one BIG step: _____

U.S. Customary Units

25. The width of two fingers: _____

26. The length of a man's shoe: _____

27. The length of one big step: _____

---

Arrange each set of units in order from shortest to longest.

28. m, cm, mm, km _____, _____, _____, _____

29. ft, mi, in., yd _____, _____, _____, _____

---

Find each equivalent measure.

30. 10 cm = _____ mm

31. 2 m = _____ cm or _____ mm

32. 5 km = _____ m

33. 2.5 cm = _____ mm

34. 1.5 m = _____ cm or _____ mm

35. 7.5 km = _____ m

36. $\frac{1}{2}$ ft = _____ in.

37. 2 ft = _____ in.

38. 3 ft = _____ in.

39. 2 yd = _____ ft

40. 10 yd = _____ ft

41. 100 yd = _____ ft

---

## L

## Liquid Measurement

*For use with Lesson 85*

Name _____

Time _____

Write the abbreviation for each of the following units.

Metric Units                    U.S. Customary Units

**1.** liter _____       **3.** ounce _____       **5.** quart _____

**2.** milliliter _____    **4.** pint _____         **6.** gallon _____

Complete each unit conversion.

Metric Conversions               U.S. Customary Conversions

**7.** 1 liter = _____ milliliters

        **8.** 1 cup   = _____ ounces

        **9.** 1 pint   = _____ ounces

      **10.** 1 quart = _____ pints

      **11.** 1 gallon = _____ quarts

Conversions between systems

**12.** 1 liter ≈ _____ quarts       **13.** 1 gallon ≈ _____ liters

Complete each statement.

**14.** One fourth of a dollar is a quarter, and one fourth of a gallon is a _____.

**15.** A two-liter bottle of soda is a little more than 2 _____ or $\frac{1}{2}$ of a _____.

**16.** "A pint's a pound the world around" means a _____ of water weighs about 1 pound.

**17.** In the United States, gasoline is sold by the _____. In other countries, gasoline is sold by the _____. A gallon of gasoline is more than _____ liters but a little less than _____ liters.

**18.** 2 cups make a _____.

**19.** 2 pints make a _____.

**20.** 2 quarts make a _____.

**21.** 2 half gallons make a _____.

**22.** A gallon of milk will fill _____ cups.

**23.** If you drink 8 cups of water each day, you drink a _____ of water.

Find each equivalent measure.

**24.** 2 liters   = _____ milliliters

**25.** 2 liters   ≈ _____ quarts

**26.** 3.78 liters = _____ milliliters

**27.** 0.5 liters = _____ milliliters

**28.** $\frac{1}{2}$ pint (1 cup)   = _____ ounces

**29.** 1 quart (2 pints) = _____ ounces

**30.** $\frac{1}{2}$ gallon   = _____ quarts

**31.** 1 gallon   = _____ pints

**32.** 2 gallons   = _____ quarts

# K Linear Measurement

*For use with Test 16*

Name _____

Time _____

---

Write the abbreviation for each of the following units.

**Metric Units**

1. millimeter _____
2. centimeter _____
3. meter _____
4. kilometer _____

**U.S. Customary Units**

5. inch _____
6. foot _____
7. yard _____
8. mile _____

---

Complete each unit conversion.

**Metric Conversions**

9. 1 centimeter = _____ millimeters
10. 1 meter = _____ millimeters
11. 1 meter = _____ centimeters
12. 1 kilometer = _____ meters

**U.S. Customary Conversions**

13. 1 foot = _____ inches
14. 1 yard = _____ inches
15. 1 yard = _____ feet
16. 1 mile = _____ feet
17. 1 mile = _____ yards

**Conversions between systems**

18. 1 inch = _____ centimeters
19. 1 mile ≈ _____ meters

20. 1 meter ≈ _____ inches
21. 1 kilometer ≈ _____ mile

---

Write an appropriate unit for each physical reference.

**Metric Units**

22. The thickness of a dime: _____
23. The width of a little finger: _____
24. The length of one BIG step: _____

**U.S. Customary Units**

25. The width of two fingers: _____
26. The length of a man's shoe: _____
27. The length of one big step: _____

---

Arrange each set of units in order from shortest to longest.

28. m, cm, mm, km _____, _____, _____, _____
29. ft, mi, in., yd _____, _____, _____, _____

---

Find each equivalent measure.

30. 10 cm = _____ mm
31. 2 m = _____ cm or _____ mm
32. 5 km = _____ m
33. 2.5 cm = _____ mm
34. 1.5 m = _____ cm or _____ mm
35. 7.5 km = _____ m

36. $\frac{1}{2}$ ft = _____ in.
37. 2 ft = _____ in.
38. 3 ft = _____ in.
39. 2 yd = _____ ft
40. 10 yd = _____ ft
41. 100 yd = _____ ft

---

© Saxon Publishers, Inc. and Stephen Hake. Reproduction prohibited.

*Saxon Math 7/6—Homeschool*

## 30 Fractions to Reduce
*For use with Lesson 86*

Name _____

Time _____

Reduce each fraction to lowest terms.

| | | | | |
|---|---|---|---|---|
| $\dfrac{2}{8} =$ | $\dfrac{4}{6} =$ | $\dfrac{6}{10} =$ | $\dfrac{2}{4} =$ | $\dfrac{6}{16} =$ |
| $\dfrac{5}{100} =$ | $\dfrac{9}{12} =$ | $\dfrac{14}{16} =$ | $\dfrac{4}{10} =$ | $\dfrac{4}{12} =$ |
| $\dfrac{2}{10} =$ | $\dfrac{3}{6} =$ | $\dfrac{25}{100} =$ | $\dfrac{3}{12} =$ | $\dfrac{4}{16} =$ |
| $\dfrac{3}{9} =$ | $\dfrac{10}{16} =$ | $\dfrac{6}{9} =$ | $\dfrac{4}{8} =$ | $\dfrac{2}{12} =$ |
| $\dfrac{6}{12} =$ | $\dfrac{2}{16} =$ | $\dfrac{8}{10} =$ | $\dfrac{2}{6} =$ | $\dfrac{75}{100} =$ |
| $\dfrac{12}{16} =$ | $\dfrac{8}{12} =$ | $\dfrac{6}{8} =$ | $\dfrac{10}{12} =$ | $\dfrac{5}{10} =$ |

## 64 Multiplication Facts
*For use with Lesson 87*

Name _____

Time _____

Multiply.

| | | | | | | | |
|---|---|---|---|---|---|---|---|
| 5<br>× 6 | 4<br>× 3 | 9<br>× 8 | 7<br>× 5 | 2<br>× 9 | 8<br>× 4 | 9<br>× 3 | 6<br>× 9 |
| 9<br>× 4 | 2<br>× 5 | 7<br>× 6 | 4<br>× 8 | 7<br>× 9 | 5<br>× 4 | 3<br>× 2 | 9<br>× 7 |
| 3<br>× 7 | 8<br>× 5 | 6<br>× 2 | 5<br>× 5 | 3<br>× 5 | 2<br>× 4 | 7<br>× 7 | 8<br>× 9 |
| 6<br>× 4 | 2<br>× 8 | 4<br>× 4 | 8<br>× 2 | 3<br>× 9 | 6<br>× 6 | 9<br>× 9 | 5<br>× 3 |
| 4<br>× 6 | 8<br>× 8 | 5<br>× 7 | 6<br>× 3 | 2<br>× 2 | 7<br>× 4 | 3<br>× 8 | 8<br>× 6 |
| 2<br>× 6 | 5<br>× 9 | 3<br>× 3 | 9<br>× 2 | 6<br>× 7 | 4<br>× 5 | 7<br>× 2 | 9<br>× 6 |
| 5<br>× 2 | 7<br>× 8 | 2<br>× 3 | 6<br>× 8 | 4<br>× 7 | 9<br>× 5 | 3<br>× 6 | 8<br>× 7 |
| 3<br>× 4 | 7<br>× 3 | 5<br>× 8 | 4<br>× 2 | 8<br>× 3 | 2<br>× 7 | 6<br>× 5 | 4<br>× 9 |

*Saxon Math 7/6—Homeschool*

| L |
|---|

**Liquid Measurement**
*For use with Lesson 88*

Name _____

Time _____

Write the abbreviation for each of the following units.

Metric Units            U.S. Customary Units

**1.** liter   _____         **3.** ounce _____         **5.** quart _____

**2.** milliliter _____       **4.** pint _____          **6.** gallon _____

Complete each unit conversion.

Metric Conversions                    U.S. Customary Conversions

**7.** 1 liter = _____ milliliters       **8.** 1 cup   = _____ ounces

                                              **9.** 1 pint   = _____ ounces

                                        **10.** 1 quart  = _____ pints

                                        **11.** 1 gallon = _____ quarts

Conversions between systems

**12.** 1 liter ≈ _____ quarts           **13.** 1 gallon ≈ _____ liters

Complete each statement.

**14.** One fourth of a dollar is a quarter, and one fourth of a gallon is a _____.

**15.** A two-liter bottle of soda is a little more than 2 _____ or $\frac{1}{2}$ of a _____.

**16.** "A pint's a pound the world around" means a _____ of water weighs about 1 pound.

**17.** In the United States, gasoline is sold by the _____. In other countries, gasoline is sold by the _____. A gallon of gasoline is more than _____ liters but a little less than _____ liters.

**18.** 2 cups make a _____.

**19.** 2 pints make a _____.

**20.** 2 quarts make a _____.

**21.** 2 half gallons make a _____.

**22.** A gallon of milk will fill _____ cups.

**23.** If you drink 8 cups of water each day, you drink a _____ of water.

Find each equivalent measure.

**24.** 2 liters    = _____ milliliters       **28.** $\frac{1}{2}$ pint (1 cup)   = _____ ounces

**25.** 2 liters    ≈ _____ quarts           **29.** 1 quart (2 pints) = _____ ounces

**26.** 3.78 liters = _____ milliliters      **30.** $\frac{1}{2}$ gallon      = _____ quarts

**27.** 0.5 liters  = _____ milliliters      **31.** 1 gallon      = _____ pints

                                                  **32.** 2 gallons     = _____ quarts

## K Linear Measurement
*For use with Lesson 89*

Name _____

Time _____

---

Write the abbreviation for each of the following units.

Metric Units

1. millimeter _____
2. centimeter _____
3. meter _____
4. kilometer _____

U.S. Customary Units

5. inch _____
6. foot _____
7. yard _____
8. mile _____

---

Complete each unit conversion.

Metric Conversions

9. 1 centimeter = _____ millimeters
10. 1 meter = _____ millimeters
11. 1 meter = _____ centimeters
12. 1 kilometer = _____ meters

U.S. Customary Conversions

13. 1 foot = _____ inches
14. 1 yard = _____ inches
15. 1 yard = _____ feet
16. 1 mile = _____ feet
17. 1 mile = _____ yards

Conversions between systems

18. 1 inch = _____ centimeters
19. 1 mile ≈ _____ meters

20. 1 meter ≈ _____ inches
21. 1 kilometer ≈ _____ mile

---

Write an appropriate unit for each physical reference.

Metric Units

22. The thickness of a dime: _____
23. The width of a little finger: _____
24. The length of one BIG step: _____

U.S. Customary Units

25. The width of two fingers: _____
26. The length of a man's shoe: _____
27. The length of one big step: _____

---

Arrange each set of units in order from shortest to longest.

28. m, cm, mm, km _____, _____, _____, _____
29. ft, mi, in., yd _____, _____, _____, _____

---

Find each equivalent measure.

30. 10 cm = _____ mm
31. 2 m = _____ cm or _____ mm
32. 5 km = _____ m
33. 2.5 cm = _____ mm
34. 1.5 m = _____ cm or _____ mm
35. 7.5 km = _____ m

36. $\frac{1}{2}$ ft = _____ in.
37. 2 ft = _____ in.
38. 3 ft = _____ in.
39. 2 yd = _____ ft
40. 10 yd = _____ ft
41. 100 yd = _____ ft

---

**J**

## 24 Mixed Numbers to Write as Improper Fractions

*For use with Lesson 90*

Name _____

Time _____

Write each mixed number as an improper fraction.

| | | | |
|---|---|---|---|
| $2\frac{1}{2} =$ | $2\frac{2}{5} =$ | $1\frac{3}{4} =$ | $2\frac{3}{4} =$ |
| $2\frac{1}{8} =$ | $1\frac{2}{3} =$ | $10\frac{1}{2} =$ | $2\frac{1}{3} =$ |
| $3\frac{1}{2} =$ | $1\frac{5}{6} =$ | $2\frac{1}{4} =$ | $1\frac{1}{8} =$ |
| $5\frac{1}{2} =$ | $1\frac{3}{8} =$ | $5\frac{1}{3} =$ | $3\frac{1}{4} =$ |
| $4\frac{1}{2} =$ | $1\frac{7}{8} =$ | $2\frac{2}{3} =$ | $3\frac{3}{10} =$ |
| $1\frac{5}{8} =$ | $3\frac{3}{4} =$ | $2\frac{3}{8} =$ | $7\frac{1}{2} =$ |

| K | **Linear Measurement** | Name _____ |
|---|---|---|
| | *For use with Test 17* | Time _____ |

Write the abbreviation for each of the following units.

Metric Units

**1.** millimeter _____

**2.** centimeter _____

**3.** meter _____

**4.** kilometer _____

U.S. Customary Units

**5.** inch _____

**6.** foot _____

**7.** yard _____

**8.** mile _____

Complete each unit conversion.

Metric Conversions

**9.** 1 centimeter = _____ millimeters

**10.** 1 meter = _____ millimeters

**11.** 1 meter = _____ centimeters

**12.** 1 kilometer = _____ meters

U.S. Customary Conversions

**13.** 1 foot = _____ inches

**14.** 1 yard = _____ inches

**15.** 1 yard = _____ feet

**16.** 1 mile = _____ feet

**17.** 1 mile = _____ yards

Conversions between systems

**18.** 1 inch = _____ centimeters

**19.** 1 mile ≈ _____ meters

**20.** 1 meter ≈ _____ inches

**21.** 1 kilometer ≈ _____ mile

Write an appropriate unit for each physical reference.

Metric Units

**22.** The thickness of a dime: _____

**23.** The width of a little finger: _____

**24.** The length of one BIG step: _____

U.S. Customary Units

**25.** The width of two fingers: _____

**26.** The length of a man's shoe: _____

**27.** The length of one big step: _____

Arrange each set of units in order from shortest to longest.

**28.** m, cm, mm, km _____, _____, _____, _____

**29.** ft, mi, in., yd _____, _____, _____, _____

Find each equivalent measure.

**30.** 10 cm = _____ mm

**31.** 2 m = _____ cm or _____ mm

**32.** 5 km = _____ m

**33.** 2.5 cm = _____ mm

**34.** 1.5 m = _____ cm or _____ mm

**35.** 7.5 km = _____ m

**36.** $\frac{1}{2}$ ft = _____ in.

**37.** 2 ft = _____ in.

**38.** 3 ft = _____ in.

**39.** 2 yd = _____ ft

**40.** 10 yd = _____ ft

**41.** 100 yd = _____ ft

**H**

## 72 Multiplication and Division Facts
*For use with Lesson 91*

Name _____

Time _____

Multiply or divide as indicated.

| | | | | | | |
|---|---|---|---|---|---|---|
| $\begin{array}{r} 5 \\ \times\ 9 \\ \hline \end{array}$ | $6\,\overline{)\,36}$ | $\begin{array}{r} 4 \\ \times\ 7 \\ \hline \end{array}$ | $8\,\overline{)\,40}$ | $\begin{array}{r} 10 \\ \times\ 3 \\ \hline \end{array}$ | $5\,\overline{)\,20}$ | $\begin{array}{r} 2 \\ \times\ 7 \\ \hline \end{array}$ | $9\,\overline{)\,27}$ |
| $9\,\overline{)\,81}$ | $\begin{array}{r} 5 \\ \times\ 7 \\ \hline \end{array}$ | $6\,\overline{)\,24}$ | $\begin{array}{r} 2 \\ \times\ 3 \\ \hline \end{array}$ | $7\,\overline{)\,42}$ | $\begin{array}{r} 7 \\ \times\ 9 \\ \hline \end{array}$ | $5\,\overline{)\,10}$ | $\begin{array}{r} 4 \\ \times\ 5 \\ \hline \end{array}$ |
| $\begin{array}{r} 2 \\ \times\ 9 \\ \hline \end{array}$ | $7\,\overline{)\,28}$ | $\begin{array}{r} 6 \\ \times\ 9 \\ \hline \end{array}$ | $4\,\overline{)\,8}$ | $\begin{array}{r} 3 \\ \times\ 3 \\ \hline \end{array}$ | $8\,\overline{)\,24}$ | $\begin{array}{r} 10 \\ \times\ 4 \\ \hline \end{array}$ | $6\,\overline{)\,12}$ |
| $9\,\overline{)\,63}$ | $\begin{array}{r} 2 \\ \times\ 5 \\ \hline \end{array}$ | $8\,\overline{)\,64}$ | $\begin{array}{r} 7 \\ \times\ 6 \\ \hline \end{array}$ | $10\,\overline{)\,100}$ | $\begin{array}{r} 3 \\ \times\ 5 \\ \hline \end{array}$ | $9\,\overline{)\,36}$ | $\begin{array}{r} 10 \\ \times\ 8 \\ \hline \end{array}$ |
| $\begin{array}{r} 4 \\ \times\ 8 \\ \hline \end{array}$ | $5\,\overline{)\,25}$ | $\begin{array}{r} 3 \\ \times\ 8 \\ \hline \end{array}$ | $4\,\overline{)\,16}$ | $\begin{array}{r} 10 \\ \times\ 5 \\ \hline \end{array}$ | $7\,\overline{)\,14}$ | $\begin{array}{r} 2 \\ \times\ 2 \\ \hline \end{array}$ | $8\,\overline{)\,48}$ |
| $9\,\overline{)\,54}$ | $\begin{array}{r} 3 \\ \times\ 6 \\ \hline \end{array}$ | $5\,\overline{)\,15}$ | $\begin{array}{r} 2 \\ \times\ 8 \\ \hline \end{array}$ | $7\,\overline{)\,49}$ | $\begin{array}{r} 4 \\ \times\ 6 \\ \hline \end{array}$ | $6\,\overline{)\,30}$ | $\begin{array}{r} 5 \\ \times\ 8 \\ \hline \end{array}$ |
| $\begin{array}{r} 10 \\ \times\ 6 \\ \hline \end{array}$ | $4\,\overline{)\,12}$ | $\begin{array}{r} 7 \\ \times\ 8 \\ \hline \end{array}$ | $6\,\overline{)\,18}$ | $\begin{array}{r} 3 \\ \times\ 4 \\ \hline \end{array}$ | $7\,\overline{)\,35}$ | $\begin{array}{r} 5 \\ \times\ 6 \\ \hline \end{array}$ | $8\,\overline{)\,32}$ |
| $8\,\overline{)\,56}$ | $\begin{array}{r} 2 \\ \times\ 4 \\ \hline \end{array}$ | $8\,\overline{)\,16}$ | $\begin{array}{r} 6 \\ \times\ 8 \\ \hline \end{array}$ | $7\,\overline{)\,21}$ | $\begin{array}{r} 8 \\ \times\ 9 \\ \hline \end{array}$ | $3\,\overline{)\,6}$ | $\begin{array}{r} 3 \\ \times\ 9 \\ \hline \end{array}$ |
| $\begin{array}{r} 10 \\ \times\ 7 \\ \hline \end{array}$ | $9\,\overline{)\,18}$ | $\begin{array}{r} 3 \\ \times\ 7 \\ \hline \end{array}$ | $9\,\overline{)\,45}$ | $\begin{array}{r} 2 \\ \times\ 6 \\ \hline \end{array}$ | $9\,\overline{)\,72}$ | $\begin{array}{r} 4 \\ \times\ 9 \\ \hline \end{array}$ | $9\,\overline{)\,90}$ |

## L  Liquid Measurement
*For use with Lesson 92*

Name _____

Time _____

Write the abbreviation for each of the following units.

Metric Units

**1.** liter   _____

**2.** milliliter _____

U.S. Customary Units

**3.** ounce _____

**4.** pint   _____

**5.** quart _____

**6.** gallon _____

---

Complete each unit conversion.

Metric Conversions

**7.** 1 liter = _____ milliliters

Conversions between systems

**12.** 1 liter ≈ _____ quarts

U.S. Customary Conversions

**8.** 1 cup   = _____ ounces

**9.** 1 pint   = _____ ounces

**10.** 1 quart   = _____ pints

**11.** 1 gallon = _____ quarts

**13.** 1 gallon ≈ _____ liters

---

Complete each statement.

**14.** One fourth of a dollar is a quarter, and one fourth of a gallon is a _____.

**15.** A two-liter bottle of soda is a little more than 2 _____ or $\frac{1}{2}$ of a _____.

**16.** "A pint's a pound the world around" means a _____ of water weighs about 1 pound.

**17.** In the United States, gasoline is sold by the _____. In other countries, gasoline is

sold by the _____. A gallon of gasoline is more than _____ liters but a little

less than _____ liters.

**18.** 2 cups make a _____.

**19.** 2 pints make a _____.

**20.** 2 quarts make a _____.

**21.** 2 half gallons make a _____.

**22.** A gallon of milk will fill _____ cups.

**23.** If you drink 8 cups of water each day, you drink a _____ of water.

---

Find each equivalent measure.

**24.** 2 liters   = _____ milliliters

**25.** 2 liters   ≈ _____ quarts

**26.** 3.78 liters = _____ milliliters

**27.** 0.5 liters = _____ milliliters

**28.** $\frac{1}{2}$ pint (1 cup)   = _____ ounces

**29.** 1 quart (2 pints) = _____ ounces

**30.** $\frac{1}{2}$ gallon   = _____ quarts

**31.** 1 gallon   = _____ pints

**32.** 2 gallons   = _____ quarts

| I | 28 Improper Fractions to Simplify | Name _____ |
|---|---|---|
| | *For use with Lesson 93* | Time _____ |

Write each improper fraction as a mixed number or a whole number.

| | | | |
|---|---|---|---|
| $\frac{5}{4} =$ | $\frac{16}{12} =$ | $\frac{12}{8} =$ | $\frac{8}{6} =$ |
| $\frac{12}{6} =$ | $\frac{12}{10} =$ | $\frac{6}{4} =$ | $\frac{20}{12} =$ |
| $\frac{5}{3} =$ | $\frac{10}{8} =$ | $\frac{25}{10} =$ | $\frac{10}{3} =$ |
| $\frac{15}{10} =$ | $\frac{3}{2} =$ | $\frac{9}{6} =$ | $\frac{7}{4} =$ |
| $\frac{18}{12} =$ | $\frac{8}{3} =$ | $\frac{15}{6} =$ | $\frac{14}{4} =$ |
| $\frac{8}{4} =$ | $\frac{10}{6} =$ | $\frac{5}{2} =$ | $\frac{21}{12} =$ |
| $\frac{15}{12} =$ | $\frac{10}{4} =$ | $\frac{15}{8} =$ | $\frac{4}{3} =$ |

# K

## Linear Measurement
*For use with Lesson 94*

Name _____

Time _____

---

Write the abbreviation for each of the following units.

Metric Units

1. millimeter _____

2. centimeter _____

3. meter _____

4. kilometer _____

U.S. Customary Units

5. inch _____

6. foot _____

7. yard _____

8. mile _____

---

Complete each unit conversion.

Metric Conversions

9. 1 centimeter = _____ millimeters

10. 1 meter = _____ millimeters

11. 1 meter = _____ centimeters

12. 1 kilometer = _____ meters

U.S. Customary Conversions

13. 1 foot = _____ inches

14. 1 yard = _____ inches

15. 1 yard = _____ feet

16. 1 mile = _____ feet

17. 1 mile = _____ yards

Conversions between systems

18. 1 inch = _____ centimeters

19. 1 mile ≈ _____ meters

20. 1 meter ≈ _____ inches

21. 1 kilometer ≈ _____ mile

---

Write an appropriate unit for each physical reference.

Metric Units

22. The thickness of a dime: _____

23. The width of a little finger: _____

24. The length of one BIG step: _____

U.S. Customary Units

25. The width of two fingers: _____

26. The length of a man's shoe: _____

27. The length of one big step: _____

---

Arrange each set of units in order from shortest to longest.

28. m, cm, mm, km _____, _____, _____, _____

29. ft, mi, in., yd _____, _____, _____, _____

---

Find each equivalent measure.

30. 10 cm = _____ mm

31. 2 m = _____ cm or _____ mm

32. 5 km = _____ m

33. 2.5 cm = _____ mm

34. 1.5 m = _____ cm or _____ mm

35. 7.5 km = _____ m

36. $\frac{1}{2}$ ft = _____ in.

37. 2 ft = _____ in.

38. 3 ft = _____ in.

39. 2 yd = _____ ft

40. 10 yd = _____ ft

41. 100 yd = _____ ft

---

*Saxon Math 7/6—Homeschool*

| **G** | **30 Fractions to Reduce** | Name _____ |
|---|---|---|
| | *For use with Lesson 95* | Time _____ |

Reduce each fraction to lowest terms.

| | | | | |
|---|---|---|---|---|
| $\frac{2}{8} =$ | $\frac{4}{6} =$ | $\frac{6}{10} =$ | $\frac{2}{4} =$ | $\frac{6}{16} =$ |
| $\frac{5}{100} =$ | $\frac{9}{12} =$ | $\frac{14}{16} =$ | $\frac{4}{10} =$ | $\frac{4}{12} =$ |
| $\frac{2}{10} =$ | $\frac{3}{6} =$ | $\frac{25}{100} =$ | $\frac{3}{12} =$ | $\frac{4}{16} =$ |
| $\frac{3}{9} =$ | $\frac{10}{16} =$ | $\frac{6}{9} =$ | $\frac{4}{8} =$ | $\frac{2}{12} =$ |
| $\frac{6}{12} =$ | $\frac{2}{16} =$ | $\frac{8}{10} =$ | $\frac{2}{6} =$ | $\frac{75}{100} =$ |
| $\frac{12}{16} =$ | $\frac{8}{12} =$ | $\frac{6}{8} =$ | $\frac{10}{12} =$ | $\frac{5}{10} =$ |

| L | **Liquid Measurement**
For use with Test 18 |

Name _____

Time _____

---

Write the abbreviation for each of the following units.

Metric Units          U.S. Customary Units

**1.** liter   _____        **3.** ounce _____        **5.** quart _____

**2.** milliliter _____      **4.** pint   _____        **6.** gallon _____

---

Complete each unit conversion.

Metric Conversions                    U.S. Customary Conversions

**7.** 1 liter = _____ milliliters        **8.** 1 cup    = _____ ounces

**9.** 1 pint    = _____ ounces

**10.** 1 quart   = _____ pints

**11.** 1 gallon = _____ quarts

Conversions between systems

**12.** 1 liter ≈ _____ quarts        **13.** 1 gallon ≈ _____ liters

---

Complete each statement.

**14.** One fourth of a dollar is a quarter, and one fourth of a gallon is a _____.

**15.** A two-liter bottle of soda is a little more than 2 _____ or $\frac{1}{2}$ of a _____.

**16.** "A pint's a pound the world around" means a _____ of water weighs about 1 pound.

**17.** In the United States, gasoline is sold by the _____. In other countries, gasoline is
sold by the _____. A gallon of gasoline is more than _____ liters but a little
less than _____ liters.

**18.** 2 cups make a _____.

**19.** 2 pints make a _____.

**20.** 2 quarts make a _____.

**21.** 2 half gallons make a _____.

**22.** A gallon of milk will fill _____ cups.

**23.** If you drink 8 cups of water each day, you drink a _____ of water.

---

Find each equivalent measure.

**24.** 2 liters    = _____ milliliters        **28.** $\frac{1}{2}$ pint (1 cup)    = _____ ounces

**25.** 2 liters    ≈ _____ quarts            **29.** 1 quart (2 pints) = _____ ounces

**26.** 3.78 liters = _____ milliliters        **30.** $\frac{1}{2}$ gallon       = _____ quarts

**27.** 0.5 liters  = _____ milliliters        **31.** 1 gallon        = _____ pints

**32.** 2 gallons       = _____ quarts

---

## D | 64 Multiplication Facts
### For use with Lesson 96

Name _____

Time _____

Multiply.

| | | | | | | | |
|---|---|---|---|---|---|---|---|
| 5<br>× 6 | 4<br>× 3 | 9<br>× 8 | 7<br>× 5 | 2<br>× 9 | 8<br>× 4 | 9<br>× 3 | 6<br>× 9 |
| 9<br>× 4 | 2<br>× 5 | 7<br>× 6 | 4<br>× 8 | 7<br>× 9 | 5<br>× 4 | 3<br>× 2 | 9<br>× 7 |
| 3<br>× 7 | 8<br>× 5 | 6<br>× 2 | 5<br>× 5 | 3<br>× 5 | 2<br>× 4 | 7<br>× 7 | 8<br>× 9 |
| 6<br>× 4 | 2<br>× 8 | 4<br>× 4 | 8<br>× 2 | 3<br>× 9 | 6<br>× 6 | 9<br>× 9 | 5<br>× 3 |
| 4<br>× 6 | 8<br>× 8 | 5<br>× 7 | 6<br>× 3 | 2<br>× 2 | 7<br>× 4 | 3<br>× 8 | 8<br>× 6 |
| 2<br>× 6 | 5<br>× 9 | 3<br>× 3 | 9<br>× 2 | 6<br>× 7 | 4<br>× 5 | 7<br>× 2 | 9<br>× 6 |
| 5<br>× 2 | 7<br>× 8 | 2<br>× 3 | 6<br>× 8 | 4<br>× 7 | 9<br>× 5 | 3<br>× 6 | 8<br>× 7 |
| 3<br>× 4 | 7<br>× 3 | 5<br>× 8 | 4<br>× 2 | 8<br>× 3 | 2<br>× 7 | 6<br>× 5 | 4<br>× 9 |

| L | **Liquid Measurement** | Name _____ |
|---|---|---|
|   | *For use with Lesson 97* | Time _____ |

Write the abbreviation for each of the following units.

Metric Units      U.S. Customary Units

**1.** liter _____      **3.** ounce _____      **5.** quart _____

**2.** milliliter _____      **4.** pint _____      **6.** gallon _____

Complete each unit conversion.

Metric Conversions      U.S. Customary Conversions

**7.** 1 liter = _____ milliliters      **8.** 1 cup = _____ ounces

**9.** 1 pint = _____ ounces

**10.** 1 quart = _____ pints

**11.** 1 gallon = _____ quarts

Conversions between systems

**12.** 1 liter ≈ _____ quarts      **13.** 1 gallon ≈ _____ liters

Complete each statement.

**14.** One fourth of a dollar is a quarter, and one fourth of a gallon is a _____.

**15.** A two-liter bottle of soda is a little more than 2 _____ or $\frac{1}{2}$ of a _____.

**16.** "A pint's a pound the world around" means a _____ of water weighs about 1 pound.

**17.** In the United States, gasoline is sold by the _____. In other countries, gasoline is sold by the _____. A gallon of gasoline is more than _____ liters but a little less than _____ liters.

**18.** 2 cups make a _____.

**19.** 2 pints make a _____.

**20.** 2 quarts make a _____.

**21.** 2 half gallons make a _____.

**22.** A gallon of milk will fill _____ cups.

**23.** If you drink 8 cups of water each day, you drink a _____ of water.

Find each equivalent measure.

**24.** 2 liters = _____ milliliters      **28.** $\frac{1}{2}$ pint (1 cup) = _____ ounces

**25.** 2 liters ≈ _____ quarts      **29.** 1 quart (2 pints) = _____ ounces

**26.** 3.78 liters = _____ milliliters      **30.** $\frac{1}{2}$ gallon = _____ quarts

**27.** 0.5 liters = _____ milliliters      **31.** 1 gallon = _____ pints

**32.** 2 gallons = _____ quarts

# FACTS PRACTICE TEST

**J**

## 24 Mixed Numbers to Write as Improper Fractions

*For use with Lesson 98*

Name _____

Time _____

Write each mixed number as an improper fraction.

| | | | |
|---|---|---|---|
| $2\frac{1}{2} =$ | $2\frac{2}{5} =$ | $1\frac{3}{4} =$ | $2\frac{3}{4} =$ |
| $2\frac{1}{8} =$ | $1\frac{2}{3} =$ | $10\frac{1}{2} =$ | $2\frac{1}{3} =$ |
| $3\frac{1}{2} =$ | $1\frac{5}{6} =$ | $2\frac{1}{4} =$ | $1\frac{1}{8} =$ |
| $5\frac{1}{2} =$ | $1\frac{3}{8} =$ | $5\frac{1}{3} =$ | $3\frac{1}{4} =$ |
| $4\frac{1}{2} =$ | $1\frac{7}{8} =$ | $2\frac{2}{3} =$ | $3\frac{3}{10} =$ |
| $1\frac{5}{8} =$ | $3\frac{3}{4} =$ | $2\frac{3}{8} =$ | $7\frac{1}{2} =$ |

*Saxon Math 7/6—Homeschool*

155

| K |

**Linear Measurement**
*For use with Lesson 99*

Name _____

Time _____

---

Write the abbreviation for each of the following units.

Metric Units

**1.** millimeter _____

**2.** centimeter _____

**3.** meter _____

**4.** kilometer _____

U.S. Customary Units

**5.** inch _____

**6.** foot _____

**7.** yard _____

**8.** mile _____

---

Complete each unit conversion.

Metric Conversions

**9.** 1 centimeter = _____ millimeters

**10.** 1 meter = _____ millimeters

**11.** 1 meter = _____ centimeters

**12.** 1 kilometer = _____ meters

U.S. Customary Conversions

**13.** 1 foot = _____ inches

**14.** 1 yard = _____ inches

**15.** 1 yard = _____ feet

**16.** 1 mile = _____ feet

**17.** 1 mile = _____ yards

Conversions between systems

**18.** 1 inch = _____ centimeters

**19.** 1 mile ≈ _____ meters

**20.** 1 meter ≈ _____ inches

**21.** 1 kilometer ≈ _____ mile

---

Write an appropriate unit for each physical reference.

Metric Units

**22.** The thickness of a dime: _____

**23.** The width of a little finger: _____

**24.** The length of one BIG step: _____

U.S. Customary Units

**25.** The width of two fingers: _____

**26.** The length of a man's shoe: _____

**27.** The length of one big step: _____

---

Arrange each set of units in order from shortest to longest.

**28.** m, cm, mm, km   _____, _____, _____, _____

**29.** ft, mi, in., yd   _____, _____, _____, _____

---

Find each equivalent measure.

**30.** 10 cm = _____ mm

**31.** 2 m = _____ cm or _____ mm

**32.** 5 km = _____ m

**33.** 2.5 cm = _____ mm

**34.** 1.5 m = _____ cm or _____ mm

**35.** 7.5 km = _____ m

**36.** $\frac{1}{2}$ ft = _____ in.

**37.** 2 ft = _____ in.

**38.** 3 ft = _____ in.

**39.** 2 yd = _____ ft

**40.** 10 yd = _____ ft

**41.** 100 yd = _____ ft

---

## 28 Improper Fractions to Simplify
*For use with Lesson 100*

Name _____

Time _____

Write each improper fraction as a mixed number or a whole number.

| | | | |
|---|---|---|---|
| $\dfrac{5}{4}$ = | $\dfrac{16}{12}$ = | $\dfrac{12}{8}$ = | $\dfrac{8}{6}$ = |
| $\dfrac{12}{6}$ = | $\dfrac{12}{10}$ = | $\dfrac{6}{4}$ = | $\dfrac{20}{12}$ = |
| $\dfrac{5}{3}$ = | $\dfrac{10}{8}$ = | $\dfrac{25}{10}$ = | $\dfrac{10}{3}$ = |
| $\dfrac{15}{10}$ = | $\dfrac{3}{2}$ = | $\dfrac{9}{6}$ = | $\dfrac{7}{4}$ = |
| $\dfrac{18}{12}$ = | $\dfrac{8}{3}$ = | $\dfrac{15}{6}$ = | $\dfrac{14}{4}$ = |
| $\dfrac{8}{4}$ = | $\dfrac{10}{6}$ = | $\dfrac{5}{2}$ = | $\dfrac{21}{12}$ = |
| $\dfrac{15}{12}$ = | $\dfrac{10}{4}$ = | $\dfrac{15}{8}$ = | $\dfrac{4}{3}$ = |

## L

### Liquid Measurement
*For use with Test 19*

Name _____

Time _____

---

Write the abbreviation for each of the following units.

Metric Units        U.S. Customary Units

**1.** liter _____      **3.** ounce _____      **5.** quart _____

**2.** milliliter _____      **4.** pint _____      **6.** gallon _____

---

Complete each unit conversion.

Metric Conversions        U.S. Customary Conversions

**7.** 1 liter = _____ milliliters      **8.** 1 cup = _____ ounces

**9.** 1 pint = _____ ounces

**10.** 1 quart = _____ pints

**11.** 1 gallon = _____ quarts

Conversions between systems

**12.** 1 liter ≈ _____ quarts      **13.** 1 gallon ≈ _____ liters

---

Complete each statement.

**14.** One fourth of a dollar is a quarter, and one fourth of a gallon is a _____.

**15.** A two-liter bottle of soda is a little more than 2 _____ or $\frac{1}{2}$ of a _____.

**16.** "A pint's a pound the world around" means a _____ of water weighs about 1 pound.

**17.** In the United States, gasoline is sold by the _____. In other countries, gasoline is sold by the _____. A gallon of gasoline is more than _____ liters but a little less than _____ liters.

**18.** 2 cups make a _____.

**19.** 2 pints make a _____.

**20.** 2 quarts make a _____.

**21.** 2 half gallons make a _____.

**22.** A gallon of milk will fill _____ cups.

**23.** If you drink 8 cups of water each day, you drink a _____ of water.

---

Find each equivalent measure.

**24.** 2 liters = _____ milliliters      **28.** $\frac{1}{2}$ pint (1 cup) = _____ ounces

**25.** 2 liters ≈ _____ quarts      **29.** 1 quart (2 pints) = _____ ounces

**26.** 3.78 liters = _____ milliliters      **30.** $\frac{1}{2}$ gallon = _____ quarts

**27.** 0.5 liters = _____ milliliters      **31.** 1 gallon = _____ pints

                                           **32.** 2 gallons = _____ quarts

---

*Saxon Math 7/6—Homeschool*

## 24 Percent-Fraction-Decimal Equivalents
*For use with Test 20*

Name _____

Time _____

Write each percent as a reduced fraction and a decimal.

| Percent | Fraction | Decimal |
|---------|----------|---------|
| 40% | | |
| 5% | | |
| 80% | | |
| 2% | | |
| 3% | | |
| 20% | | |
| 25% | | |
| 60% | | |
| 1% | | |
| 90% | | |
| 75% | | |
| 10% | | |

| Percent | Fraction | Decimal |
|---------|----------|---------|
| 70% | | |
| 4% | | |
| 100% | | |
| 30% | | |
| 50% | | |
| $12\frac{1}{2}\%$ | | |
| $37\frac{1}{2}\%$ | | |
| $62\frac{1}{2}\%$ | | |
| $87\frac{1}{2}\%$ | | |
| $33\frac{1}{3}\%$ | | Rounds to 0.333 |
| $66\frac{2}{3}\%$ | | Rounds to 0.667 |
| $16\frac{2}{3}\%$ | | Rounds to 0.167 |

## K  Linear Measurement
*For use with Lesson 106*

Name _____

Time _____

---

Write the abbreviation for each of the following units.

Metric Units

   **1.** millimeter _____

   **2.** centimeter _____

   **3.** meter     _____

   **4.** kilometer _____

U.S. Customary Units

   **5.** inch _____

   **6.** foot _____

   **7.** yard _____

   **8.** mile _____

---

Complete each unit conversion.

Metric Conversions

   **9.** 1 centimeter = _____ millimeters

 **10.** 1 meter    = _____ millimeters

 **11.** 1 meter    = _____ centimeters

 **12.** 1 kilometer = _____ meters

U.S. Customary Conversions

 **13.** 1 foot = _____ inches

 **14.** 1 yard = _____ inches

 **15.** 1 yard = _____ feet

 **16.** 1 mile = _____ feet

 **17.** 1 mile = _____ yards

Conversions between systems

 **18.** 1 inch = _____ centimeters

 **19.** 1 mile ≈ _____ meters

 **20.** 1 meter    ≈ _____ inches

 **21.** 1 kilometer ≈ _____ mile

---

Write an appropriate unit for each physical reference.

Metric Units

 **22.** The thickness of a dime: _____

 **23.** The width of a little finger: _____

 **24.** The length of one BIG step: _____

U.S. Customary Units

 **25.** The width of two fingers: _____

 **26.** The length of a man's shoe: _____

 **27.** The length of one big step: _____

---

Arrange each set of units in order from shortest to longest.

 **28.** m, cm, mm, km _____, _____, _____, _____

 **29.** ft, mi, in., yd _____, _____, _____, _____

---

Find each equivalent measure.

 **30.** 10 cm = _____ mm

 **31.** 2 m     = _____ cm or _____ mm

 **32.** 5 km    = _____ m

 **33.** 2.5 cm = _____ mm

 **34.** 1.5 m   = _____ cm or _____ mm

 **35.** 7.5 km = _____ m

 **36.** $\frac{1}{2}$ ft     = _____ in.

 **37.** 2 ft     = _____ in.

 **38.** 3 ft     = _____ in.

 **39.** 2 yd    = _____ ft

 **40.** 10 yd   = _____ ft

 **41.** 100 yd = _____ ft

## M

### 24 Percent-Fraction-Decimal Equivalents
*For use with Lesson 107*

Name _____

Time _____

Write each percent as a reduced fraction and a decimal.

| Percent | Fraction | Decimal |
|---------|----------|---------|
| 40% | | |
| 5% | | |
| 80% | | |
| 2% | | |
| 3% | | |
| 20% | | |
| 25% | | |
| 60% | | |
| 1% | | |
| 90% | | |
| 75% | | |
| 10% | | |

| Percent | Fraction | Decimal |
|---------|----------|---------|
| 70% | | |
| 4% | | |
| 100% | | |
| 30% | | |
| 50% | | |
| $12\frac{1}{2}\%$ | | |
| $37\frac{1}{2}\%$ | | |
| $62\frac{1}{2}\%$ | | |
| $87\frac{1}{2}\%$ | | |
| $33\frac{1}{3}\%$ | | Rounds to 0.333 |
| $66\frac{2}{3}\%$ | | Rounds to 0.667 |
| $16\frac{2}{3}\%$ | | Rounds to 0.167 |

# M | 24 Percent-Fraction-Decimal Equivalents

*For use with Lesson 108*

Name _____

Time _____

Write each percent as a reduced fraction and a decimal.

| Percent | Fraction | Decimal |
|---------|----------|---------|
| 40% | | |
| 5% | | |
| 80% | | |
| 2% | | |
| 3% | | |
| 20% | | |
| 25% | | |
| 60% | | |
| 1% | | |
| 90% | | |
| 75% | | |
| 10% | | |

| Percent | Fraction | Decimal |
|---------|----------|---------|
| 70% | | |
| 4% | | |
| 100% | | |
| 30% | | |
| 50% | | |
| $12\frac{1}{2}$% | | |
| $37\frac{1}{2}$% | | |
| $62\frac{1}{2}$% | | |
| $87\frac{1}{2}$% | | |
| $33\frac{1}{3}$% | | Rounds to 0.333 |
| $66\frac{2}{3}$% | | Rounds to 0.667 |
| $16\frac{2}{3}$% | | Rounds to 0.167 |

## M

## 24 Percent-Fraction-Decimal Equivalents

*For use with Test 21*

Name _____

Time _____

Write each percent as a reduced fraction and a decimal.

| Percent | Fraction | Decimal |
|---------|----------|---------|
| 40% | | |
| 5% | | |
| 80% | | |
| 2% | | |
| 3% | | |
| 20% | | |
| 25% | | |
| 60% | | |
| 1% | | |
| 90% | | |
| 75% | | |
| 10% | | |

| Percent | Fraction | Decimal |
|---------|----------|---------|
| 70% | | |
| 4% | | |
| 100% | | |
| 30% | | |
| 50% | | |
| $12\frac{1}{2}\%$ | | |
| $37\frac{1}{2}\%$ | | |
| $62\frac{1}{2}\%$ | | |
| $87\frac{1}{2}\%$ | | |
| $33\frac{1}{3}\%$ | | Rounds to 0.333 |
| $66\frac{2}{3}\%$ | | Rounds to 0.667 |
| $16\frac{2}{3}\%$ | | Rounds to 0.167 |

## 17

## Scale Model
*For use with Investigation 11*

Art by Dan Shippey of Delta 7 Studios

### Instructions

**Step 1:** Carefully cut out each piece on both pages. Label the back of each piece with its corresponding letter label.

**Step 2:** Glue (B), (D), (E), and (G) in a loop as instructed on each piece. Carefully cut (F) along the dashed line and glue as indicated.

**Step 3:** Attach (A) to (B) by folding the tabs on (A) down and gluing them to (B). Attach (C) to (D) and (H) to (G) in the same manner.

**Step 4:** Attach (G) to (F) by folding out the black tabs on (G) and gluing them to (F). Attach (B) to (C) in the same manner. The tabs on (B) should line up with the white rectangles on (C).

**Step 5:** Locate the tab labeled "X" on (D). Glue this tab to the corresponding rectangle labeled "X" on (E). Then glue the rest of the tabs on (D) to (E). Attach (F) to (E) in the same manner.

### Assembly Diagram

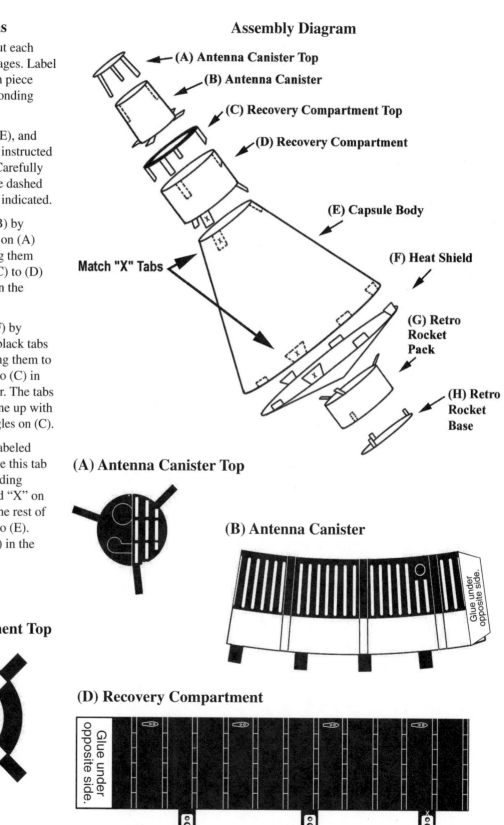

- (A) Antenna Canister Top
- (B) Antenna Canister
- (C) Recovery Compartment Top
- (D) Recovery Compartment
- (E) Capsule Body
- (F) Heat Shield
- (G) Retro Rocket Pack
- (H) Retro Rocket Base

Match "X" Tabs

**(A) Antenna Canister Top**

**(B) Antenna Canister**

Glue under opposite side.

**(C) Recovery Compartment Top**

**(D) Recovery Compartment**

Glue under opposite side.

## Scale Model
*For use with Investigation 11*

Art by Dan Shippey of Delta 7 Studios

**(E) Capsule Body**

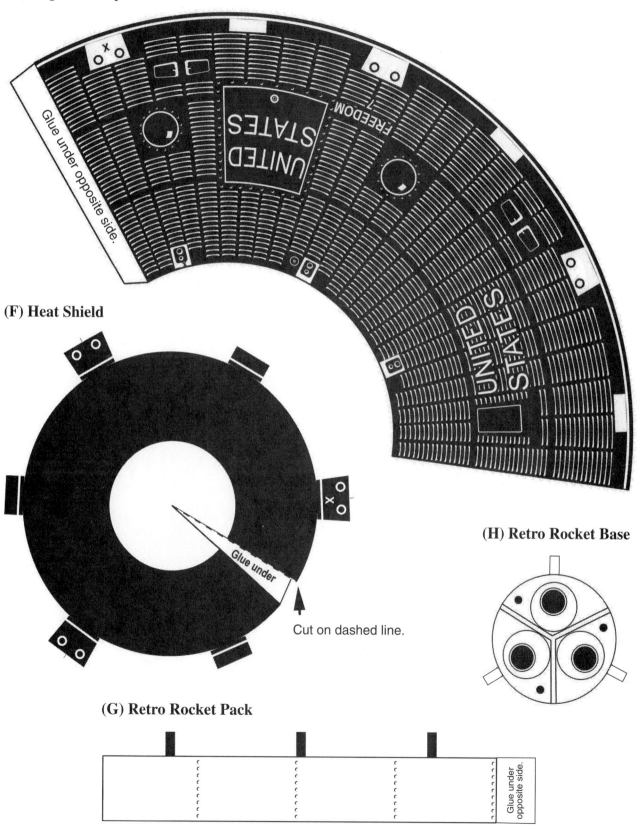

**(F) Heat Shield**

**(H) Retro Rocket Base**

Cut on dashed line.

**(G) Retro Rocket Pack**

## N

### Measurement Facts
*For use with Lesson 111*

Name _____

Time _____

1. Draw a segment 1 cm long.

2. Draw a segment 1 in. long.

3. One inch is how many centimeters? _____

4. Which is longer, 1 km or 1 mi? _____

5. Which is longer, 1 km or $\frac{1}{2}$ mi? _____

6. How many ounces are in a pound? _____

7. How many pounds are in a ton? _____

8. A dollar bill has a mass of about one _____.

9. Your math book has a mass of about one _____.

10. On earth a kilogram mass weighs about _____ pounds.

11. A metric ton is _____ kilograms.

12. On earth a metric ton weighs about _____ pounds.

13. The earth rotates on its axis once in a _____.

14. The earth revolves around the sun once in a _____.

Abbreviations:

15. milligram _____

16. gram _____

17. kilogram _____

18. _____ C

19. ounce _____

20. pound _____

21. ton _____

22. _____ F

Equivalents:

23. 1 gram = _____ milligrams

24. 1 kilogram = _____ grams

25. $\frac{1}{2}$ ton = _____ pounds

26. _____ days = a common year

27. _____ days = a leap year

28. _____ weeks ≈ a year

29. _____ years = a decade

30. _____ years = a century

31. _____ years = a millennium

How many days are in

32. Jan. _____

33. Feb. _____ or _____

34. Mar. _____

35. Apr. _____

36. May _____

37. June _____

38. July _____

39. Aug. _____

40. Sept. _____

41. Oct. _____

42. Nov. _____

43. Dec. _____

Write the indicated temperatures.

44. Water boils _____ °F

45. _____ °C

46. Normal body temperature _____ °F

47. _____ °C

Cool room temperature 68 °F

48. _____ °C

49. Water freezes _____ °F

50. _____ °C

51. A cubic container 1 cm on each edge has a volume of one _____ and can hold one _____ of water, which has a mass of one _____.

52. A cubic container 10 cm on each edge has a volume of _____ cubic centimeters and can hold one _____ of water which has a mass of one _____.

1 cm
1 cm
1 cm

10 cm

10 cm

10 cm

*Saxon Math 7/6—Homeschool*

| M | 24 Percent-Fraction-Decimal Equivalents |
|---|---|

*For use with Lesson 112*

Name _____

Time _____

Write each percent as a reduced fraction and a decimal.

| Percent | Fraction | Decimal |
|---------|----------|---------|
| 40% | | |
| 5% | | |
| 80% | | |
| 2% | | |
| 3% | | |
| 20% | | |
| 25% | | |
| 60% | | |
| 1% | | |
| 90% | | |
| 75% | | |
| 10% | | |

| Percent | Fraction | Decimal |
|---------|----------|---------|
| 70% | | |
| 4% | | |
| 100% | | |
| 30% | | |
| 50% | | |
| $12\frac{1}{2}\%$ | | |
| $37\frac{1}{2}\%$ | | |
| $62\frac{1}{2}\%$ | | |
| $87\frac{1}{2}\%$ | | |
| $33\frac{1}{3}\%$ | | Rounds to 0.333 |
| $66\frac{2}{3}\%$ | | Rounds to 0.667 |
| $16\frac{2}{3}\%$ | | Rounds to 0.167 |

## N Measurement Facts
*For use with Lesson 113*

Name _____

Time _____

1. Draw a segment 1 cm long.

2. Draw a segment 1 in. long.

3. One inch is how many centimeters? _____

4. Which is longer, 1 km or 1 mi? _____

5. Which is longer, 1 km or $\frac{1}{2}$ mi? _____

6. How many ounces are in a pound? _____

7. How many pounds are in a ton? _____

8. A dollar bill has a mass of about one _____.

9. Your math book has a mass of about one _____.

10. On earth a kilogram mass weighs about _____ pounds.

11. A metric ton is _____ kilograms.

12. On earth a metric ton weighs about _____ pounds.

13. The earth rotates on its axis once in a _____.

14. The earth revolves around the sun once in a _____.

Abbreviations:

15. milligram _____

16. gram _____

17. kilogram _____

18. _____ C

19. ounce _____

20. pound _____

21. ton _____

22. _____ F

Equivalents:

23. 1 gram = _____ milligrams

24. 1 kilogram = _____ grams

25. $\frac{1}{2}$ ton = _____ pounds

26. _____ days = a common year

27. _____ days = a leap year

28. _____ weeks ≈ a year

29. _____ years = a decade

30. _____ years = a century

31. _____ years = a millennium

How many days are in

32. Jan. _____

33. Feb. _____ or _____

34. Mar. _____

35. Apr. _____

36. May _____

37. June _____

38. July _____

39. Aug. _____

40. Sept. _____

41. Oct. _____

42. Nov. _____

43. Dec. _____

Write the indicated temperatures.

44. Water boils _____ °F

45. _____ °C

46. Normal body temperature _____ °F

47. _____ °C

Cool room temperature 68 °F

48. _____ °C

49. Water freezes _____ °F

50. _____ °C

51. A cubic container 1 cm on each edge has a volume of one _____ _____ and can hold one _____ of water, which has a mass of one _____.

52. A cubic container 10 cm on each edge has a volume of _____ cubic centimeters and can hold one _____ of water which has a mass of one _____.

1 cm

1 cm   1 cm

10 cm

10 cm

10 cm

| M |

## 24 Percent-Fraction-Decimal Equivalents
*For use with Lesson 114*

Name _____

Time _____

Write each percent as a reduced fraction and a decimal.

| Percent | Fraction | Decimal |
|---------|----------|---------|
| 40% | | |
| 5% | | |
| 80% | | |
| 2% | | |
| 3% | | |
| 20% | | |
| 25% | | |
| 60% | | |
| 1% | | |
| 90% | | |
| 75% | | |
| 10% | | |

| Percent | Fraction | Decimal |
|---------|----------|---------|
| 70% | | |
| 4% | | |
| 100% | | |
| 30% | | |
| 50% | | |
| $12\frac{1}{2}\%$ | | |
| $37\frac{1}{2}\%$ | | |
| $62\frac{1}{2}\%$ | | |
| $87\frac{1}{2}\%$ | | |
| $33\frac{1}{3}\%$ | | Rounds to 0.333 |
| $66\frac{2}{3}\%$ | | Rounds to 0.667 |
| $16\frac{2}{3}\%$ | | Rounds to 0.167 |

*Saxon Math 7/6—Homeschool*

## N  Measurement Facts
*For use with Lesson 115*

Name _____

Time _____

1. Draw a segment 1 cm long.

2. Draw a segment 1 in. long.

3. One inch is how many centimeters? _____

4. Which is longer, 1 km or 1 mi? _____

5. Which is longer, 1 km or $\frac{1}{2}$ mi? _____

6. How many ounces are in a pound? _____

7. How many pounds are in a ton? _____

8. A dollar bill has a mass of about one _____.

9. Your math book has a mass of about one _____.

10. On earth a kilogram mass weighs about _____ pounds.

11. A metric ton is _____ kilograms.

12. On earth a metric ton weighs about _____ pounds.

13. The earth rotates on its axis once in a _____.

14. The earth revolves around the sun once in a _____.

Abbreviations:

15. milligram _____

16. gram _____

17. kilogram _____

18. _____ C

19. ounce _____

20. pound _____

21. ton _____

22. _____ F

Equivalents:

23. 1 gram = _____ milligrams

24. 1 kilogram = _____ grams

25. $\frac{1}{2}$ ton = _____ pounds

26. _____ days = a common year

27. _____ days = a leap year

28. _____ weeks ≈ a year

29. _____ years = a decade

30. _____ years = a century

31. _____ years = a millennium

How many days are in

32. Jan. _____

33. Feb. _____ or _____

34. Mar. _____

35. Apr. _____

36. May _____

37. June _____

38. July _____

39. Aug. _____

40. Sept. _____

41. Oct. _____

42. Nov. _____

43. Dec. _____

Write the indicated temperatures.

44. Water boils _____ °F

45. _____ °C

46. Normal body temperature _____ °F

47. _____ °C

Cool room temperature   68   °F

48. _____ °C

49. Water freezes _____ °F

50. _____ °C

51. A cubic container 1 cm on each edge has a volume of one _____ _____ and can hold one _____ of water, which has a mass of one _____.

52. A cubic container 10 cm on each edge has a volume of _____ cubic centimeters and can hold one _____ of water which has a mass of one _____.

1 cm
1 cm
1 cm

10 cm

10 cm

10 cm

| M |
|---|

## 24 Percent-Fraction-Decimal Equivalents
*For use with Test 22*

Name _____

Time _____

Write each percent as a reduced fraction and a decimal.

| Percent | Fraction | Decimal |
|---------|----------|---------|
| 40% | | |
| 5% | | |
| 80% | | |
| 2% | | |
| 3% | | |
| 20% | | |
| 25% | | |
| 60% | | |
| 1% | | |
| 90% | | |
| 75% | | |
| 10% | | |

| Percent | Fraction | Decimal |
|---------|----------|---------|
| 70% | | |
| 4% | | |
| 100% | | |
| 30% | | |
| 50% | | |
| $12\frac{1}{2}\%$ | | |
| $37\frac{1}{2}\%$ | | |
| $62\frac{1}{2}\%$ | | |
| $87\frac{1}{2}\%$ | | |
| $33\frac{1}{3}\%$ | | Rounds to 0.333 |
| $66\frac{2}{3}\%$ | | Rounds to 0.667 |
| $16\frac{2}{3}\%$ | | Rounds to 0.167 |

# M

## 24 Percent-Fraction-Decimal Equivalents

*For use with Lesson 116*

Name _____

Time _____

Write each percent as a reduced fraction and a decimal.

| Percent | Fraction | Decimal |
|---------|----------|---------|
| 40% | | |
| 5% | | |
| 80% | | |
| 2% | | |
| 3% | | |
| 20% | | |
| 25% | | |
| 60% | | |
| 1% | | |
| 90% | | |
| 75% | | |
| 10% | | |

| Percent | Fraction | Decimal |
|---------|----------|---------|
| 70% | | |
| 4% | | |
| 100% | | |
| 30% | | |
| 50% | | |
| $12\frac{1}{2}$% | | |
| $37\frac{1}{2}$% | | |
| $62\frac{1}{2}$% | | |
| $87\frac{1}{2}$% | | |
| $33\frac{1}{3}$% | | Rounds to 0.333 |
| $66\frac{2}{3}$% | | Rounds to 0.667 |
| $16\frac{2}{3}$% | | Rounds to 0.167 |

## N  Measurement Facts
*For use with Lesson 117*

Name _____

Time _____

1. Draw a segment 1 cm long.

2. Draw a segment 1 in. long.

3. One inch is how many centimeters? _____

4. Which is longer, 1 km or 1 mi? _____

5. Which is longer, 1 km or $\frac{1}{2}$ mi? _____

6. How many ounces are in a pound? _____

7. How many pounds are in a ton? _____

8. A dollar bill has a mass of about one _____.

9. Your math book has a mass of about one _____.

10. On earth a kilogram mass weighs about _____ pounds.

11. A metric ton is _____ kilograms.

12. On earth a metric ton weighs about _____ pounds.

13. The earth rotates on its axis once in a _____.

14. The earth revolves around the sun once in a _____.

Abbreviations:

15. milligram _____

16. gram _____

17. kilogram _____

18. _____ C

19. ounce _____

20. pound _____

21. ton _____

22. _____ F

Equivalents:

23. 1 gram = _____ milligrams

24. 1 kilogram = _____ grams

25. $\frac{1}{2}$ ton = _____ pounds

26. _____ days = a common year

27. _____ days = a leap year

28. _____ weeks ≈ a year

29. _____ years = a decade

30. _____ years = a century

31. _____ years = a millennium

How many days are in

32. Jan. _____

33. Feb. _____ or _____

34. Mar. _____

35. Apr. _____

36. May _____

37. June _____

38. July _____

39. Aug. _____

40. Sept. _____

41. Oct. _____

42. Nov. _____

43. Dec. _____

Write the indicated temperatures.

44. Water boils _____ °F

45. _____ °C

46. Normal body temperature _____ °F

Cool room temperature 68 °F

47. _____ °C

48. _____ °C

49. Water freezes _____ °F

50. _____ °C

1 cm  1 cm  1 cm

51. A cubic container 1 cm on each edge has a volume of one _____ _____ and can hold one _____ of water, which has a mass of one _____.

52. A cubic container 10 cm on each edge has a volume of _____ cubic centimeters and can hold one _____ of water which has a mass of one _____.

10 cm

10 cm

10 cm

**M**

## 24 Percent-Fraction-
## Decimal Equivalents
*For use with Lesson 118*

Name _____

Time _____

Write each percent as a reduced fraction and a decimal.

| Percent | Fraction | Decimal |
|---------|----------|---------|
| 40% | | |
| 5% | | |
| 80% | | |
| 2% | | |
| 3% | | |
| 20% | | |
| 25% | | |
| 60% | | |
| 1% | | |
| 90% | | |
| 75% | | |
| 10% | | |

| Percent | Fraction | Decimal |
|---------|----------|---------|
| 70% | | |
| 4% | | |
| 100% | | |
| 30% | | |
| 50% | | |
| $12\frac{1}{2}\%$ | | |
| $37\frac{1}{2}\%$ | | |
| $62\frac{1}{2}\%$ | | |
| $87\frac{1}{2}\%$ | | |
| $33\frac{1}{3}\%$ | | Rounds to 0.333 |
| $66\frac{2}{3}\%$ | | Rounds to 0.667 |
| $16\frac{2}{3}\%$ | | Rounds to 0.167 |

**N** **Measurement Facts**
*For use with Lesson 119*

Name _____

Time _____

1. Draw a segment 1 cm long.

2. Draw a segment 1 in. long.

3. One inch is how many centimeters? _____

4. Which is longer, 1 km or 1 mi? _____

5. Which is longer, 1 km or $\frac{1}{2}$ mi? _____

6. How many ounces are in a pound? _____

7. How many pounds are in a ton? _____

8. A dollar bill has a mass of about one _____.

9. Your math book has a mass of about one _____.

10. On earth a kilogram mass weighs about _____ pounds.

11. A metric ton is _____ kilograms.

12. On earth a metric ton weighs about _____ pounds.

13. The earth rotates on its axis once in a _____.

14. The earth revolves around the sun once in a _____.

Abbreviations:

15. milligram _____

16. gram _____

17. kilogram _____

18. _____ C

19. ounce _____

20. pound _____

21. ton _____

22. _____ F

Equivalents:

23. 1 gram = _____ milligrams

24. 1 kilogram = _____ grams

25. $\frac{1}{2}$ ton = _____ pounds

26. _____ days = a common year

27. _____ days = a leap year

28. _____ weeks ≈ a year

29. _____ years = a decade

30. _____ years = a century

31. _____ years = a millennium

How many days are in

32. Jan. _____

33. Feb. _____ or _____

34. Mar. _____

35. Apr. _____

36. May _____

37. June _____

38. July _____

39. Aug. _____

40. Sept. _____

41. Oct. _____

42. Nov. _____

43. Dec. _____

Write the indicated temperatures.

44. Water boils _____ °F

45. _____ °C

46. Normal body temperature _____ °F

47. _____ °C

Cool room temperature 68 °F

48. _____ °C

49. Water freezes _____ °F

50. _____ °C

°F    °C

51. A cubic container 1 cm on each edge has a volume of one _____ _____ and can hold one _____ of water, which has a mass of one _____.

1 cm
1 cm
1 cm

52. A cubic container 10 cm on each edge has a volume of _____ cubic centimeters and can hold one _____ of water which has a mass of one _____.

10 cm

10 cm

10 cm

*Saxon Math 7/6—Homeschool*

# M

## 24 Percent-Fraction-Decimal Equivalents
*For use with Lesson 120*

Name _____

Time _____

Write each percent as a reduced fraction and a decimal.

| Percent | Fraction | Decimal |
|---------|----------|---------|
| 40% | | |
| 5% | | |
| 80% | | |
| 2% | | |
| 3% | | |
| 20% | | |
| 25% | | |
| 60% | | |
| 1% | | |
| 90% | | |
| 75% | | |
| 10% | | |

| Percent | Fraction | Decimal |
|---------|----------|---------|
| 70% | | |
| 4% | | |
| 100% | | |
| 30% | | |
| 50% | | |
| $12\frac{1}{2}\%$ | | |
| $37\frac{1}{2}\%$ | | |
| $62\frac{1}{2}\%$ | | |
| $87\frac{1}{2}\%$ | | |
| $33\frac{1}{3}\%$ | | Rounds to 0.333 |
| $66\frac{2}{3}\%$ | | Rounds to 0.667 |
| $16\frac{2}{3}\%$ | | Rounds to 0.167 |

**N** | **Measurement Facts**
*For use with Test 23*

Name _____

Time _____

1. Draw a segment 1 cm long.

2. Draw a segment 1 in. long.

3. One inch is how many centimeters? _____

4. Which is longer, 1 km or 1 mi? _____

5. Which is longer, 1 km or $\frac{1}{2}$ mi? _____

6. How many ounces are in a pound? _____

7. How many pounds are in a ton? _____

8. A dollar bill has a mass of about one _____.

9. Your math book has a mass of about one _____.

10. On earth a kilogram mass weighs about _____ pounds.

11. A metric ton is _____ kilograms.

12. On earth a metric ton weighs about _____ pounds.

13. The earth rotates on its axis once in a _____.

14. The earth revolves around the sun once in a _____.

Abbreviations:

15. milligram _____

16. gram _____

17. kilogram _____

18. _____ C

19. ounce _____

20. pound _____

21. ton _____

22. _____ F

Equivalents:

23. 1 gram = _____ milligrams

24. 1 kilogram = _____ grams

25. $\frac{1}{2}$ ton = _____ pounds

26. _____ days = a common year

27. _____ days = a leap year

28. _____ weeks ≈ a year

29. _____ years = a decade

30. _____ years = a century

31. _____ years = a millennium

How many days are in

32. Jan. _____

33. Feb. _____ or _____

34. Mar. _____

35. Apr. _____

36. May _____

37. June _____

38. July _____

39. Aug. _____

40. Sept. _____

41. Oct. _____

42. Nov. _____

43. Dec. _____

Write the indicated temperatures.

44. Water boils _____ °F

45. _____ °C

46. Normal body temperature _____ °F

Cool room temperature 68 °F

47. _____ °C

48. _____ °C

49. Water freezes _____ °F

50. _____ °C

51. A cubic container 1 cm on each edge has a volume of one _____ _____ and can hold one _____ of water, which has a mass of one _____.

52. A cubic container 10 cm on each edge has a volume of _____ cubic centimeters and can hold one _____ of water which has a mass of one _____.

**Tetrahedron and Cube Patterns**
*For use with Investigation 12*

**Tetrahedron** (four-faced polyhedron)

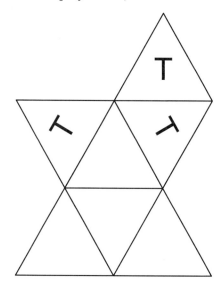

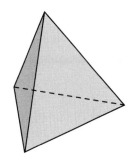

**Cube**

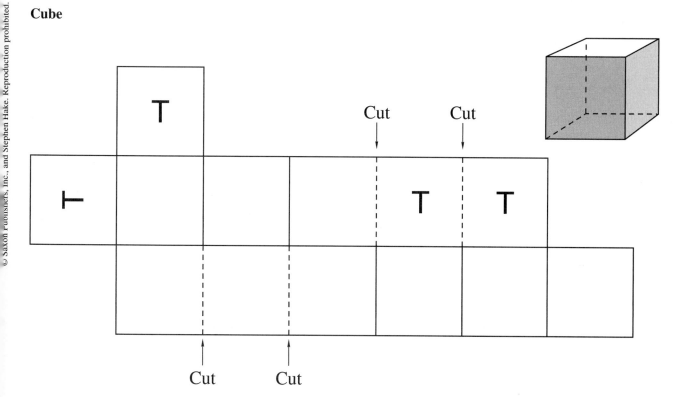

## Octahedron and Dodecahedron Patterns
*For use with Investigation 12*

**Octahedron**

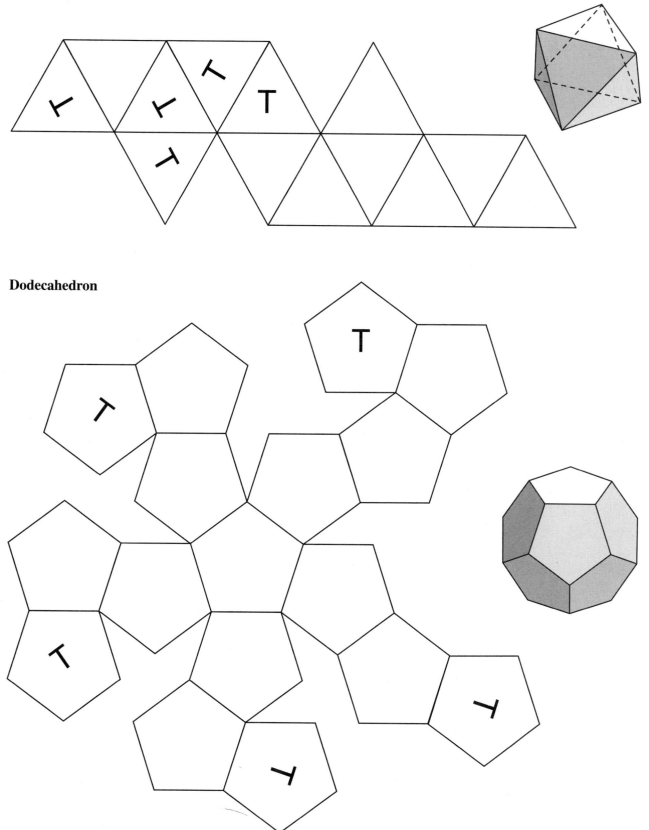

**Dodecahedron**

## 21 | Icosahedron Pattern
*For use with Investigation 12*

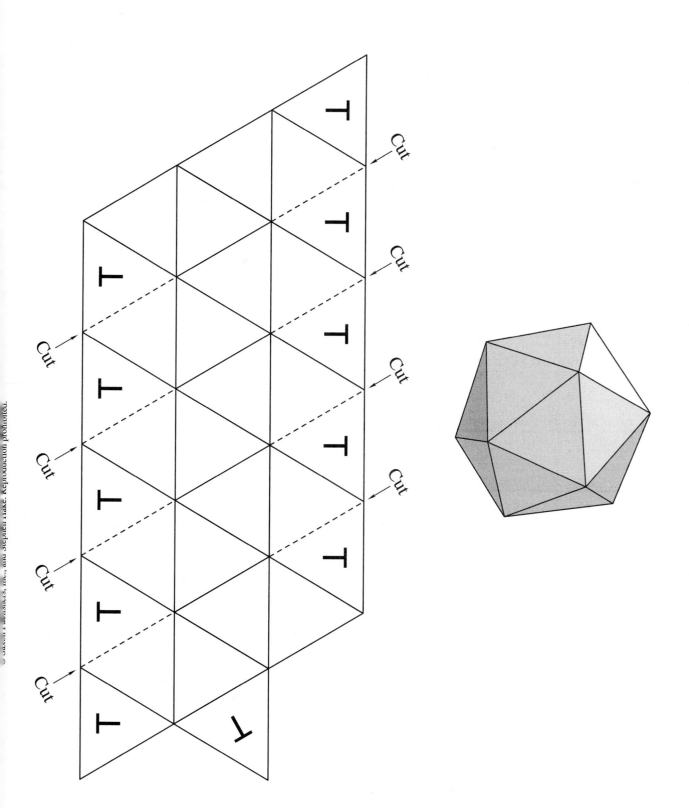

# *Tests*

A test should be given after every fifth lesson, beginning after Lesson 10. The testing schedule is explained in greater detail on the back of this page.

On test days, allow five minutes for your student to take the Facts Practice Test indicated at the top of the test. Then administer the cumulative test specified by the Testing Schedule. You might wish also to provide your student with a photocopy of Recording Form E. This form is designed to provide an organized space for your student to show his or her work. *Note:* The textbook should not be used during the test.

Solutions to the test problems are located in the *Saxon Math 7/6—Homeschool Solutions Manual.* For detailed information on appropriate test-grading strategies, please refer to the preface in the *Saxon Math 7/6—Homeschool* textbook.

# Testing Schedule

| Test to be administered | Covers material through | Give after |
|:---:|:---:|:---:|
| Test 1 | Lesson 5 | Lesson 10 ✓ |
| Test 2 | Lesson 10 | Lesson 15 ✓ |
| Test 3 | Lesson 15 | Lesson 20 ✓ |
| Test 4 | Lesson 20 | Lesson 25 ✓ |
| Test 5 | Lesson 25 | Lesson 30 ✓ |
| Test 6 | Lesson 30 | Lesson 35 ✓ |
| Test 7 | Lesson 35 | Lesson 40 ✓ |
| Test 8 | Lesson 40 | Lesson 45 ✓ |
| Test 9 | Lesson 45 | Lesson 50 ✓ |
| Test 10 | Lesson 50 | Lesson 55 ✓ |
| Test 11 | Lesson 55 | Lesson 60 ✓ |
| Test 12 | Lesson 60 | Lesson 65 ✓ |
| Test 13 | Lesson 65 | Lesson 70 ✓ |
| Test 14 | Lesson 70 | Lesson 75 ✓ |
| Test 15 | Lesson 75 | Lesson 80 ✓ |
| Test 16 | Lesson 80 | Lesson 85 ✓ |
| Test 17 | Lesson 85 | Lesson 90 ✓ |
| Test 18 | Lesson 90 | Lesson 95 ✓ |
| Test 19 | Lesson 95 | Lesson 100 ✓ |
| Test 20 | Lesson 100 | Lesson 105 ✓ |
| Test 21 | Lesson 105 | Lesson 110 ✓ |
| Test 22 | Lesson 110 | Lesson 115 ✓ |
| Test 23 | Lesson 115 | Lesson 120 |

# *Recording Forms*

The five optional recording forms in this section may be photocopied to provide the quantities needed by you and your student.

**Recording Form A:  Facts Practice**
> This form helps your student track his or her performances on Facts Practice Tests throughout the year.

**Recording Form B:  Lesson Worksheet**
> This single-sided form is designed to be used with daily lessons. It contains a checklist of the daily lesson routine as well as answer blanks for the Warm-Up and Lesson Practice.

**Recording Form C:  Mixed Practice Solutions**
> This double-sided form provides a framework for your student to show his or her work on the Mixed Practices. It has a grid background and partitions for recording the solutions to thirty problems.

**Recording Form D:  Scorecard**
> This form is designed to help you and your student track scores on daily assignments and cumulative tests.

**Recording Form E:  Test Solutions**
> This double-sided form provides a framework for your student to show his or her work on the tests. It has a grid background and partitions for recording the solutions to twenty problems.

## B | Lesson Worksheet

*Show all necessary work. Please be neat.*

Name _____

Date _____

Lesson _____

### Warm-Up
- ☐ Facts Practice
- ☐ Mental Math
- ☐ Problem Solving

### Review
- ☐ Homework Check
- ☐ Error Correction

### Instruction
- ☐ Lesson
- ☐ Lesson Practice
- ☐ Mixed Practice

### Facts Practice

| Test: | Time: | Score: |
|---|---|---|

### Mental Math

| a. | b. | c. | d. | e. | f. |
|---|---|---|---|---|---|
| g. | h. | i. | j. | k. | l. |

### Problem Solving

Strategies:
(Check any you use.)

- ☐ Make a chart, graph, or list.
- ☐ Guess and check (trial and error).
- ☐ Use logical reasoning.
- ☐ Act it out.  ☐ Draw a diagram.
- ☐ Make it simpler.  ☐ Draw a picture.
- ☐ Work backward.  ☐ Find a pattern.

### Lesson Practice

| a. | b. | c. |
|---|---|---|
| d. | e. | f. |
| g. | h. | i. |
| j. | k. | l. |

*Saxon Math 7/6—Homeschool*

# RECORDING FORM

## C | Mixed Practice Solutions

*Show all necessary work. Please be neat.*

Name _____

Date _____

Lesson _____

2.

3.

5.

6.

8.

9.

11.

12.

14.

15.

*Saxon Math 7/6—Homeschool*

**16.**

**17.**

**18.**

**19.**

**20.**

**21.**

**22.**

**23.**

**24.**

**25.**

**26.**

**27.**

**28.**

**29.**

**30.**

| D | Scorecard | | Name _____ |

| Date | Lesson or Test | Score | Date | Lesson or Test | Score | Date | Lesson or Test | Score | Date | Lesson or Test | Score |
|---|---|---|---|---|---|---|---|---|---|---|---|
| | | | | | | | | | | | |
| | | | | | | | | | | | |
| | | | | | | | | | | | |
| | | | | | | | | | | | |
| | | | | | | | | | | | |
| | | | | | | | | | | | |
| | | | | | | | | | | | |
| | | | | | | | | | | | |
| | | | | | | | | | | | |
| | | | | | | | | | | | |
| | | | | | | | | | | | |
| | | | | | | | | | | | |
| | | | | | | | | | | | |
| | | | | | | | | | | | |
| | | | | | | | | | | | |
| | | | | | | | | | | | |
| | | | | | | | | | | | |
| | | | | | | | | | | | |